Fundamental Generation Systems

Digital Sciences Set

coordinated by
Abdelkhalak El Hami

Volume 4

Fundamental Generation Systems

Computer Science and Artificial Consciousness, the Informational Field of Generation of the Universe, the Sixth Sense of Living Beings

Alain Cardon
Abdelkhalak El Hami

WILEY

First published 2023 in Great Britain and the United States by ISTE Ltd and John Wiley & Sons, Inc.

ISTE Ltd
27-37 St George's Road
London SW19 4EU
UK

www.iste.co.uk

John Wiley & Sons, Inc.
111 River Street
Hoboken, NJ 07030
USA

www.wiley.com

Any opinions, findings, and conclusions or recommendations expressed in this material are those of the author(s), contributor(s) or editor(s) and do not necessarily reflect the views of ISTE Group.

Library of Congress Control Number: 2023930941

British Library Cataloguing-in-Publication Data
A CIP record for this book is available from the British Library
ISBN 978-1-78630-873-3

Contents

Preface   xi

Introduction   xv

Chapter 1. Systems and their Designs   1

  1.1. System modeling   1
    1.1.1. Traditional systems   2
    1.1.2. Complex systems   2
    1.1.3. Systems of systems   3
  1.2. Autonomous systems   4
  1.3. Multi-agent systems   6
  1.4. Organizations and systems   7
  1.5. The problem of modeling an autonomous system   8
  1.6. Agents and multi-agent systems   9
    1.6.1. The weak notion of an agent   10
    1.6.2. The strong notion of an agent   11
    1.6.3. Cognitive agents and reactive agents   11
    1.6.4. Multi-agent systems   13
    1.6.5. MAS with reactive agents   14
    1.6.6. MAS with cognitive agents   14

Chapter 2. Reliability of Autonomous Systems   17

  2.1. Introduction   17
  2.2. Dependability of a system   18
    2.2.1. General concepts   18
    2.2.2. Failure and repair rates   21
    2.2.3. Average estimators   24
    2.2.4. Some methodological tools   27

2.3. Reliability diagram   28
    2.3.1. Series system   29
    2.3.2. Parallel system   30
    2.3.3. Mixed system   31
    2.3.4. More complex systems.   32
    2.3.5. Fault tree   33
2.4. Reliability networks.   34
    2.4.1. Partial graph associated with a subset of components   34
    2.4.2. Reliability network and structure function   35
    2.4.3. Properties of reliability networks.   35
    2.4.4. Length and width of a reliability network   36
    2.4.5. Equivalence between structure function and reliability network . .   36
    2.4.6. Construction and simplification of reliability networks   36

Chapter 3. Artificial Intelligence, Communication Systems and
Artificial Consciousness   39

3.1. Introduction.   39
3.2. Evolution of computer science   39
3.3. Evolution of artificial intelligence   43
3.4. Radical evolution of computing and AI towards fully communicating
systems   47
3.5. The computer representation of an artificial consciousness   53

Chapter 4. The Informational Substrate of the Universe and the
Organizational Law   63

4.1. Introduction.   63
4.2. The fundamental principles of the informational model of generation
of the Universe   63
    4.2.1. First fundamental concept: the foundation of the Universe by a
    generative information system   64
    4.2.2. Second fundamental concept: informational character of the
    Universe with its substrate   65
    4.2.3. Third fundamental concept: the organizational law of generation
    of the Universe   67
    4.2.4. Fourth fundamental concept: the informational energy of the
    substrate of the Universe   69
4.3. The notion of generating information in the Universe.   71
    4.3.1. The information field of a component.   72
    4.3.2. The informational activity component and its information
    envelope   73
    4.3.3. Fifth fundamental concept: self-control in the organization of the
    Universe   74

4.3.4. Generative information   75
4.3.5. Organizational tendency of the Universe and informational
envelopes   76

Chapter 5. The Informational Interpretation of Living Things   81

5.1. Introduction.   81
5.2. Origin of living things with bifurcation of the organizational law. . . .   82
5.2.1. Sixth fundamental concept: the production of living things
on Earth   82
5.2.2. Bifurcation of the organizational law   83
5.2.3. Seventh fundamental concept: the central rule of organizational
law in living things.   84
5.2.4. The principle of action of the organizational law for the generation
of living things   85
5.2.5. The life span of a living organism   87
5.2.6. The unification of the informational envelope with the membrane.   88
5.2.7. The creation of sexual reproduction   89
5.2.8. Reasons for the production of new organisms   90
5.3. The informational action of reproduction of living things   91
5.3.1. The fundamental rule of the organizational law that formed
living beings.   93
5.3.2. Morphological patterns   94
5.3.3. The influence of an external morphological pattern on a
living organism   95
5.3.4. Brain formation and sensory comprehension.   97
5.3.5. The external organizational attractors hypothesis   99
5.3.6. The possibility of predicting the future of any situation
in progress.   99
5.4. The human species in the organizational evolution of living things. . .   100
5.4.1. Creation of *Homo sapiens* as a result of very strong evolution . . .   101
5.4.2. Organizational action of the formation of the human brain   101
5.4.3. The importance of informational links between groups of humans.   103
5.4.4. Power of group participation in humans   104

Chapter 6. The Interpretation of Neuronal Aggregates   107

6.1. Introduction.   107
6.1.1. Form of a thought.   108
6.1.2. Constructivist definition of the notion of mental representation . .   108
6.2. The systemic layer and the regulators including the
informational regulator.   112
6.2.1. Systemic layer of the psychic system   112
6.2.2. Regulators   113

6.2.3. The voluntary choice of the operated aim in the psychic system . .   113
6.2.4. The aggregate–regulator rule of co-activity   114
6.2.5. Morphological role of regulators.   116
6.2.6. Process of intentionally producing a representation on a
desired theme   121
6.2.7. The organizational immersion regulator in an
informational state   124

**Chapter 7. The Sense of Informational Comprehension of Living
Organisms: The Sixth Sense**   125

7.1. Introduction.   125
7.2. The five usual senses and the use of the informational substrate   126
7.2.1. The case of animals.   127
7.2.2. The case of plants.   128
7.3. The sense of informational comprehension or the sixth sense   129
7.3.1. The implicit communication process of the sixth sense   130
7.3.2. The voluntary process of informational communication   131
7.3.3. The solicitation of the sixth sense and the comprehension of the
informational substrate   135
7.3.4. The cognitive and sensitive interpretation of the information
forms received   139
7.4. Common use of the sixth sense   142
7.4.1. Common use of the sixth sense in animals and humans   142
7.4.2. The action of the sixth sense for hypnosis, the power of
magnetism and meditation   147
7.4.3. The notion of premonition and the sixth sense.   151
7.4.4. Thoughts and the safeguarding of the world   153

Conclusion.   155

Appendices   157

Appendix 1. Binomial Distribution   159

Appendix 2. Geometric Distribution   161

Appendix 3. Poisson Distribution   163

Appendix 4. Exponential Distribution   165

Appendix 5. Normal Distribution.   167

Appendix 6. Lognormal Distribution   171

Appendix 7. Weibull Distribution   175

Appendix 8. Pareto Distribution   179

Appendix 9. Distribution of Extreme Values.   183

Appendix 10. Asymptotic Distributions   185

References.   191

Index   195

Preface

There are many different ways of generating representations. This includes representations generated by living beings while comprehending reality in order to act; representations generated by the Universe during its extensive unfolding, creating physical elements and living beings; and, the direct representation of elements through an animal's sixth sense. To this list we must now add the creation of artificial consciousness, which generates representations that resemble the mental representations of humans. These representations allow robotic systems to communicate directly with each other.

In our research, we have stated that the Universe, with all its material elements, in addition to the living organisms on Earth, was generated and then organized on a strictly informational substrate. This informational substrate – which is a set of virtual processes and relations – founds space, time and the elements of the Universe. The Universe is founded on a set of informational fields that constitute its foundation and deployment. It is not a set of elements that randomly structure themselves on available nothingness forming space, but a system that is conceived and generated, strictly autonomous and in self-organized development.

Today, computer science, along with artificial intelligence, is moving towards total communication between all humans and between all computerized systems, which is a transposition of the sixth sense in the field of sophisticated techniques. In addition, there is the generation of artificial consciousness systems that allow robotic systems to communicate directly with one another, generating shared or even common artificial mental representations, which goes beyond the sixth sense in animals.

In Chapter 1, we present systems and their designs. We explore the modeling of systems, traditional systems, complex systems, systems of systems, autonomous systems, multi-agent systems and finally organisms and systems.

Chapter 2 presents the reliability of autonomous systems. It highlights the reliability of a system, its general concepts, the failure and repair rates, the average estimators and some methodological tools for modeling the reliability of systems (serial, parallel, mixed) and finally the complex systems.

Chapter 3 presents computer science, artificial intelligence, communication systems and artificial consciousness. Today, computer science is a core discipline in science and society due to the innumerable uses of software that constantly communicate. We will trace its history, show how significant artificial intelligence has become and witness the move towards a distributed, highly communicating and autonomous artificial consciousness. These continuous communications between humans through computerized systems come from a tendency to communicate, which human technology has strongly developed and whose source comes from the informational substratum of the Universe, which is essentially communicational.

Chapter 4 presents the informational substrate of the Universe and the organizational law. We present the Universe as an organizational system generating space and physical elements. The Universe was created by a very specific soliciting element, which produced informational elements that created space and structured elements in a continuous way with spatial and temporal stability. We have therefore shown that the Universe is an organized emergence with informational components, whose role is to aggregate into physical elements on an *informational substratum* that carries out a *self-control incentive*. This generation is achieved by following an organizational law that operates at the level of the informational substratum. The model we have presented allows us to consider the Universe as the continuous generation of a self-organizing system that creates its space, the matter of its physical elements, based on specific and absolutely continuous informational communications.

Chapter 5 presents the informational interpretation of the living. In particular, the origin of life with a ramification of the organizational law, the informational action of reproduction of life with morphological patterns and the human species in the organizational evolution of life. We can consider that all living organisms on Earth are stabilized structures of informational fields that are part of a general multi-scale organization, localized in the geographical areas that make up the terrestrial ecosystem, in order to behave and develop there by generating multiple species. It is thus put forward that all living things on Earth are immersed in an informational domain that comes from the informational substratum of the Universe, which is based on a considerable set of informational fields that make it possible to structure the organisms by incentives to their organizations, with control over their generation. All living organisms are therefore the very finely organized material realizations of the reifications of these informational fields.

Chapter 6 is devoted to the representation of human consciousness with its senses. It presents the interpretation of neuronal aggregates, the systemic layer and the regulators, including the informational regulator. The neuronal system operates in terms of parallel production of multiple neuronal signals which, by their associations and aggregations, form a very complex whole that can be interpreted as a structure of dynamic forms that combine with one another. This is a structure made up of activities and informational exchanges that carry sensitive and cognitive indications at a certain level.

Chapter 7 is devoted to the informational understanding of living organisms: the sixth sense. This sixth sense is unlike the five usual senses, because it is a global and organizational comprehension of elements that are outside the understanding of the five usual senses. We will show how this sense is a comprehension of elements of the informational network that forms the organization of all elements on Earth, by developing what this informational comprehension of communicating informational envelopes is, which is completely different from a visual comprehension of a world of objects with measurable positions in space. In addition, we will explain the notion of magnetism used by healers.

February 2023

Introduction

We therefore assume that the informational substratum of the Universe is generated from a generative component that produces innumerable informational components, each one also producing others, according to a general organizational law that enables the generation and expansion of our whole Universe. Each created component either becomes an element with neutrality of activity, and is therefore an element of space, or is an element with permanence of activity and is a basic quantum element. The active elements can generate others according to the organizational law and the context. These transformations are therefore subject to an organizational law that allows the organized generation of the Universe with all its elements. The Universe has been created from a generating component with informational energy and, by continuous generation, has produced informational components that reproduce themselves according to the underlying organizational law, and constitute space and quantum particles. According to the general tendency defined by the organizational law that drives the substratum, these particles will aggregate to form molecules and then physical material elements, in a specific time-frame, which will be the speed of the unfolding of the Universe in generation. The physical Universe is thus an extensive unfolding, which is self-organized by an informational system.

All the generated informational components are elements whose role is to form aggregations, following the organizational law that drives the formation of the Universe. The structured physical elements created will each have an informational envelope, a specific informational field, which will indicate their specificities and current states, and which will connect them to the other elements by informational links that are fields, in order to continue the aggregations and proceed towards the formation of massive elements. The notion of materiality is thus based on the informational substratum of the Universe that makes all the elements exist, producing structured aggregates that have structural stability.

The organizational law is the cause of the evolutionary living on Earth, which benefited from favorable conditions: stability and the existence of water. All living organisms have an informational envelope enabling their generation, organization and organically coherent functioning. This envelope is the reason for the formation of new species in an informational management of reproduction. This physical informational envelope is the synthesis of the informational envelopes of each organ, which allows coherence in the functioning network of any living organism. The general envelope of the organism describes the general state of the organism, whether it is in good or bad condition. Living beings, including all animals, use this envelope to directly and immediately understand the situation of certain other living beings that they know and are interested in: they make use of their sixth sense.

Human thought is complex. Placing ourselves within the framework of constructing an ideal multi-character representation that is generated and felt, in addition to the use of language, and under multiple simultaneous constraints, to think is to construct a series of mental representations on themes. Constructions and reconstructions produce the sensation of thinking by their existence and characters. We can define what this system is capable of producing as a set of forms that can be manipulated, while considering the depth and richness of the experience that allows us to define them, with tendencies expressing the ability to abstract, formulate, open up to the knowable external world and the five senses, which are always usable.

Today, it is generally agreed that animals are endowed with a sixth sense, which is an ability leading to the production of sensitive and conceptual representations, expressing the situations and movements of particular individuals or the approach and state of known physical places, but which are not apprehensible by the five usual senses. We will develop a theory presenting the function of this sixth sense, as well as its origin, something which has not yet been done, and we will call this sixth sense the "sense of informational comprehension". This sense will be conceived as an ability to use the informational substratum of the Universe, with a specific cerebral domain which exists in all animals. There is therefore a very particular cerebral domain in the brain, which makes it possible to communicate through the informational links of the substrate. This sixth sense is of considerable importance in the living world, as it causes a tendency in all these organisms towards living in groups and forming associations by systematically using the informational envelopes of the participants in the groups. Using this sense, animals are able to directly understand the general state of other organisms and their spatial location, by accessing their informational envelopes. In humans, it is their tendency to form organized groups. It is therefore a very important ability for a human being at the social level to voluntarily and socially develop their use of this sixth sense, as it enables each human to carry out direct informational communication with a great many other humans, by understanding their organizational states, their proximities or their distances, and thus using this informational immersion to move towards ethical sharing, peace and the organizational unification of all living beings.

Present-day computer science, along with artificial intelligence, is moving towards total communication between all humans and between all computerized systems, which is a transposition of the sixth sense in the field of sophisticated techniques. In addition, there is the generation of artificial consciousness systems that allow robotic systems to communicate directly with one another, generating shared or even common artificial mental representations, which go beyond the sixth sense in animals.

1

Systems and their Designs

1.1. System modeling

A system is normally designed to provide services. It consists of hardware, software and human resources to meet a specific, clearly defined need. The history of science is full of such systems. Their manufacturing methods have evolved over time as a result of the experiences acquired, the evolution of technologies and the modeling approaches. Various notions can intervene in the description of a system. They concern its components and their groupings, their interactions and the interactions with the environment of the system.

Generally, the notion of a system implies interdependent entities, whose functionalities are fully specified. The system is clearly defined according to an equational and functional approach, in a top-down or bottom-up iterative process. It is top-down when the approach is analytical and allows each part to be broken down into sub-parts, which are sub-systems in their own right. The reverse iterative process – bottom-up – is instead oriented towards the construction of sub-systems from the more basic ones. The implementation of the system and its possible evolution are predetermined in a narrow and precise field, the functionalities being able to relate to various and varied fields: electrical, electronic, data-processing, mechanical, etc.

With the evolution of systems and the progress made in information and communication technologies, we can observe a trend of building large systems with an increasing number of strongly interconnected elements and handling very large amounts of data.

There are different types of systems, but we will only consider two separate classes here: traditional systems and complex systems.

1.1.1. *Traditional systems*

The so-called individual or traditional systems are those whose inputs/outputs are completely specified, in the sense that everything has been designated for them at their origins. They form the vast majority of the systems we encounter. This is the case of a management application, scientific calculation or musical creation, for example. The elements constituting these systems are determined to carry out a specific process for which the system was formatted. This processing produces actions or results to be exploited, which is the purpose of the system. Even though a system is operational but still evolving, as long as it has a project manager, it is a traditional system. Everything is framed for such systems. An example would be an ATM. All the conditions of use must be clearly defined and supervised to allow it to function normally, in order to meet the demands of the customers and the bank. Operation in a degraded mode or in the case of unforeseen events must also be considered.

The development of computer networks contributes to the evolution of these traditional systems, by increasing the possibilities of exploiting their resources and by enriching their possibilities of interaction. This also contributes to the complexity of these systems, but without changing their basic nature: they remain traditional. Service-oriented architecture (SOA) is one example. The development of cloud computing from the perspective of the services offered is an illustration. The accumulation of systems for the accumulation of services offers systems that also remain traditional as long as the services obtained can be deduced from the sum of the services of the systems that make them up. The integration of systems to produce new desired services produces a new traditional system, by its functional description. Malfunction situations are normally also managed.

1.1.2. *Complex systems*

The literature describes many types of systems, with a particular interest in complex systems, mainly because of the non-predictability of their behavior. They generally concern fields where multidisciplinarity is expressed (economy, neuroscience, insect societies, etc.).

Authors generally agree to the idea of defining a complex system as a system that is composed of a large number of interacting entities, whose behavior cannot be deduced from the behavior of its parts. Hence, the concept of emergence: the complex system has an emergent behavior that cannot be deduced from any of the systems that constitute it. It is not the large size of a system that makes it complex: if its parts have been designed, structured and interact in a known way, then it is not complex. However, a non-complex system becomes complex as soon as a human is part of it.

Complex systems have many behavioral characteristics that are sought after for study: self-organization, emergence, non-determinism, etc. The common approach to the study of complex systems is simulation, because it allows us to get an idea, even though only partially, of the system's behavior. Complex systems show an autonomy of behavior that we will specify later, by linking it to the notion of proactivity.

Any information system that includes both functional elements and elements that consider human actions and decisions, as well as manage multiple points of view, is a complex system where the components are placed at multiple scales that can vary from one to another.

1.1.3. *Systems of systems*

The notion of system of systems (SoS) (Jamshidi 2008) has been introduced without being outlined by a clear definition. Several approaches are suggested in the literature. At first glance, the notion implies the existence of several systems that operate together (Zeigler and Sarjoughian 2013). From this notion of SoS, we exclude all that brings us back to the case of traditional systems, which can thus be reduced to centralized management, as is the case for a family of systems. Among the SoS approaches, let us mention the case of the super system comprising complex independent elements that interact for a common goal, or that of a large-scale system of competing and distributed systems.

The most common notion of SoS (sometimes called complex SoS) (Maier 1999) implies a set of systems, each one having been defined for its own services and for which it is managed, but without justifying its presence in the global system. The global system must also exhibit emergent behavior. An SoS thus benefits from the activity of these systems, in order to build its own. There can be a large number of these systems, but that can also change. Each system can join or leave the global system at any time. This sheds light on the absence of a predefined objective for the global system, as well as the difference in the mode of control. In other words, the overall goal of an SoS may not have been defined a priori.

The scalability characteristic of SoS through integration of other systems can be due to reasons such as rapid technological change or budgetary reasons. These give an SoS the ability to be "quickly" augmented or reduced by parts (Cardon and Itmi 2016). This point of view shows that the engineering of SoS cannot be conventionally carried out, as for traditional systems, according to a top-down or bottom-up process.

This approach highlights a particular architecture, with a functioning that implies coordination/regulation and a basis that is manifested by an orientation towards one or several goals. This raises various questions concerning the notion of autonomy, the reason for organization in autonomous systems, the notion of coherence of behavior and the orientation of the activity, as well as the control of such a system.

Distributed simulation is a way to approach SoS. It is similar to the simulation of a peer-to-peer system, but necessary elements are required for an understanding of the emergent behavior (see Figure 1.1).

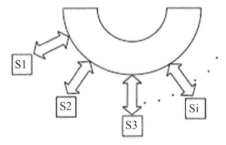

Figure 1.1. *A peer-to-peer organization around a network*

1.2. Autonomous systems

The notion of an autonomous system (in the field of robotics) implies a system that has the capacity to act by itself, to carry out actions that are necessary for the realization of predefined objectives responding to stimuli which, in robotics, come from sensors. Various approaches to the notion of autonomy exist in the literature, because the ability to act by oneself can take on different characteristics, from the activity of an automaton to that of a living being, as well as through those of a system that has the ability to act by learning.

Along with the notion of an autonomous system, which represents an advance on systems that are not advanced, there is the notion of intelligent control, which represents an advance on the notion of control. Intelligent control involves algorithms, linguistics and mathematics applied to systems and processes (Cardon and Itmi 2016), and for hierarchical systems, control is described in a three-level model, which is widely referenced in the literature. We will mention it briefly in what follows and refer the reader to the original article for more details. In short, the three levels are:

– the organizational level;

– the coordination level;

– the execution level.

It should be noted that in the first level, the imitation of human functions is sought with an interest in analytical approaches. We can make the following observations about this approach:

– The proposed model is a hierarchical (top-down) model, defining a machine that is subject to the dictates of the organizational level (see how the information feedback is done).

– It is an approach that heavily relies on calculation and ignores the work on knowledge. As a result, processing is done in a "closed world" and does not seem to us to adapt to multidisciplinarity.

– The definitions associated with the various levels construct this difference: for example, the first two levels do not consider the notions of organization and emergence.

– The integration of two systems does not seem possible to us with the approach of Sarano (2017). What about the integration of a proactive system? The notion of proactivity is completely absent. Working on a priori knowledge is to already have some control, a proactive element cannot be well controlled.

– Another important point is the absence of the notion of point of view, although it is a major one in our case. We start from the hypothesis that knowledge depends on the point of view, which relativizes it. Knowledge is therefore subjective, and we do not assume absolute truth.

In this work, we propose a model of autonomous systems that has been directly inspired by the living world and which differs from the model above. Our approach will show that we do not address the same problems as strictly analytical approaches.

In order for the system to behave like an autonomous organism, it must have an architecture that is composed of elements considered as artificial organs and, above all, at the elementary level, it must be designed by computer elements that also have minimal autonomy, which are sensitive to their environment and which are modified by the simple fact of being activated, of functioning.

1.3. Multi-agent systems

A multi-agent system (MAS) is a system composed of a set of agents forming an organization, in other words, an identified system that reorganizes itself through its actions and the relations between its elements, that conforms and re-conforms itself to carry out its action on the environment. Unlike the systems developed in artificial intelligence, which simulate certain capacities of human reasoning in a certain field, based on knowledge structures where an inference-type reasoning mechanism operates, MAS are conceived and implemented as a set of agents which interact according to modes of cooperation, competition and negotiation, and which therefore continuously conform their behavioral organization to identify the most effective form each time.

A MAS is characterized by the following:

– Each agent has limited information and problem-solving capabilities. It has knowledge and appreciation that are only partial or local to the general problem that the MAS must address and solve.

– There is no global and centralized control of the MAS, which is the main point.

– The data of the system are decentralized and taken by some system interface agents, by managing the distribution and temporality problems.

– The calculation of the solution of the problem that the system must solve at each phase of interpellation, and thus the functioning of the MAS, is done by managing the coordination of all the agents, and this is achieved in an asynchronous way in order to identify a small set of agents, by emergence, that will perform the action and act on the problem.

Moreover, a MAS can be seen as a set of agents, located in an environment that is composed of other agents and objects different from the agents. Indeed, the agents use objects of the environment (object in the strictly functional and computing sense of the term), which are simply reactive elements providing information and producing functional actions. The agents have a capacity to interpret the information given by the methods of these objects, as well as the behavior of the other agents, taking the time to do so. These agents thus use the objects and communicate with other agents in order to achieve their objectives. In the system that we define, this makes it possible to distinguish all of the information to be comprehended, which will be produced systematically by the objects whose role it is to do so, from their analyses and conceptual interpretations at multiple levels, which will be the purpose of the organization of the agents.

1.4. Organizations and systems

In all that follows, we will focus on open systems, in other words, those that interact with their environment. Such systems are understood as a set of elements in relations, organizing the action of their interdependent elements and producing actions on their environments. Such systems are therefore both the set of their elements and the continuous relations allowing them to exist and act on their environments.

In biology, an organism is the set of organs of a living being. An organ is a term in biology that designates a set of tissues, whose activity performs a specific physiological function or small set of functions. This organ is an element of a biological system that performs all the functions of a specific domain. Organs and their relations are represented by anatomical diagrams, anatomical charts, etc., and all these particular biological systems are placed in the unified whole that constitutes and forms the living organism. We can then identify the organism with the living being.

A comparison can be made between certain artificial systems and natural organisms, by examining their compositions and the underlying relations.

The relations between the elements of a system can be considered as information manipulation. For this, we will consider two levels in a system:

– the level of physical elements, formed by basic elements and aggregates of such elements;

– the level of exchange and manipulation of information between physical elements.

We therefore adopt an approach that transposes the fundamental nature of living organisms into the domain of artificial systems. This will involve an original design approach and the use of very specific elements.

1.5. The problem of modeling an autonomous system

Artificial corporeality is a set of distributed electronic and computer elements with precise functions that are managed locally by computer processors, but with an overall unification of all these actions, making it possible to give meaning to their links and actions by coordinating them in a continuous way.

An artificial organ will be a specific element, formed by the union of a particular electronic system activating electromechanical elements and a computer control system unifying these elements, making it possible to represent their precise functionalities in order to use them in very coordinated ways. This organ is arranged in a corporality made of multiple other organs, and it will be managed as a highly coactive element with all other elements considered as organs.

To design an artificial organism, it will be necessary to rely on two major concepts that will lead to the definition of its complex architecture:

– One of these concepts will involve the organization of the physical components of the system, which will be considered as its organs and which will form its highly organized corporality.

– The other major concept will be to design an interpretation system that continuously manages the system's behavioral states, interpreting and processing all the information it comprehends, using all its knowledge. This interpretation system will make it possible to continuously generate sequences of its own representations, with intention, while deriving what it comprehends, what it conceives, what it represents to itself and what it wants, thus committing it to desired and continuously interpreted actions.

It is quite clear that it is a question of providing the system with a representation sequence generator, so that it can express its intentions, its desires and its wishes, by experiencing sensations. The design of such a system that fully uses its corporality and comprehends itself as an organism is the key to the current notion of autonomy.

Through the emergence of representation sequences, the interpretation system, which is the key to system autonomy, will generate what is comprehended and desired by the system at any moment. Such a system, strictly at computer level, will be able to represent an artificial proto-self. The knowledge to be represented in such a system is very specific. We propose the use of swarms of active software agents that will have to be controlled and directed towards emergences that are producing representations of what is comprehended. We thus also propose an incentive self-control, which has not been developed so far.

This is what a truly autonomous system is: it does not limit itself to the use of multiple pieces of knowledge to produce predefined responses to more or less complex situations, but instead cognitively and significantly interprets the reality it comprehends, in order to deploy and fully establish itself. The computer system that will immerse the entirety of this distributed physical system will be the major system of the autonomous artificial organism, and we will present its precise architecture.

1.6. Agents and multi-agent systems

The agent concept is used in many different fields. We therefore encounter definitions that differ according to the field in which the agent is conceived, such as the notion of an economic agent, which is used in economics to represent the behavior of a selfish human player, and does not relate to the field of computer science. More precisely, in our field, an agent is:

> An active, autonomous entity capable of performing specific tasks. This definition comes from A. Newell's rational agent, where the level of knowledge is above the symbolic level and where the considerations represented by rational agents are proper knowledge, goals, and means of action and communication (Newell 1982).

Specifically, an agent is:

– an intelligent entity, acting rationally and intentionally with respect to its goals and the current state of its knowledge;

– a high-level, slaved entity, acting continuously and autonomously in an environment where processes take place and other agents exist.

On another note, M. Wooldridge and N.R. Jennings introduced the strong and weak notion of an agent, in order to specify bounds in the concept (Wooldridge and Jennings 1994).

1.6.1. *The weak notion of an agent*

An agent that qualifies as weak must have the following characteristics:

– It must be able to act without the intervention of a third party (human or agent) and must be able to control its own actions, as well as its internal state, using predefined rules.

– It must have a certain sociality, in other words, be able to interact with other agents (software or human) when the situation requires it, in order to complete its tasks or to help these agents to accomplish theirs.

– It must be proactive, in other words, exhibit opportunistic behavior, and be able to make decisions.

1.6.2. *The strong notion of an agent*

In addition to the capabilities of agents responding to the weak notion described above, those responding to the strong notion have the following, according to the two authors:

– "beliefs": what the agent knows about its environment and interprets;

– "desires": what the agent's goals are, defined according to motivations;

– "intentions": in order to realize its desires, the agent uses actions that manifest its intentions.

This strong notion makes these agents appear as complex systems which are actually autonomous, and not as usual software agents, which are the constitutive elements of a system which will be complex. The characters defined are absolutely non-trivial, inspired by human psyche, which artificial intelligence specialists do not really know how to describe well, based on the traditional formalism of knowledge. We will not follow this notion of a strong agent, and we will instead orient ourselves towards systems handling organizations that are composed of a very large number of weak agents, by assuming that abilities in beliefs, desires and intentions can only exist at the level of the organization of the set of agents, which will act in order to make forms appear by emergence.

1.6.3. *Cognitive agents and reactive agents*

Initially, in computer science, there were two ways of looking at agents. The first way was known as "cognitive", which considers the agent as an intelligent entity capable of solving certain problems by itself. Each of these agents then possesses a certain reduced knowledge base, plans and goals, in order to carry out and plan its tasks, making these entities cooperate and communicate with each other, which can be described as "intelligent". To this end, many research works are adopting this character in the agents by relying on sociological works, in order to solve coordination problems between actors.

The second way of seeing agents is called "reactive". The latter considers that the intelligent behavior of the system must emerge from the interaction between various agent behaviors, much simpler behaviors. According to this point of view, we build agents without using complex cognitive representations or fine reasoning mechanisms in their structures. These agents only have multiple reaction mechanisms to the events they perceive.

It is now clearly considered that each agent has cognitive capacities, capacities that are limited, but which are effective because they are thoroughly implemented and defined at the design stage by rules and meta-rules in the agents' structures. The problem of linking agents, of their interactions and of freeing the hegemonic agents remains the core problem that must be solved, in order to make the most efficient system behavior possible emerge from the set of active agents, and to have it well adapted to the current situation. Through this approach, we will not focus the problems on the notion of a single agent, but on the notion of organizations of agents, each of which will be formed of many or very many agents, and whose interactions will be used and controlled. We then get to the notion of a MAS, which is a well-organized set of agents carrying out a certain number of actions, constituting the behavior of a system.

However, we can give a minimal definition of agents in the constructivist spirit of the field of system modeling. In this regard, we can quote the properties that agents should have as design entities, according to J. Ferber (1999) an agent:

– is able to act in a planned manner in its environment;

– offers skills and services;

– has its own resources;

– is able to perceive its environment, in a limited way, by only having a partial representation of this environment;

– can communicate directly with other agents through links called bridging relationships;

– is active with individual objectives and a function of satisfaction, or even survival, of the objectives that it seeks to reach or to optimize;

– displays behavior tending to satisfy its objectives, taking into account its resources and competencies, and according to its perception and the communications it receives.

1.6.4. *Multi-agent systems*

A MAS is thus a system composed of a set of agents forming an organization, that is, an identified system that reorganizes itself by its actions and the relations between its elements, which conforms and re-conforms itself to realize its action on the environment. Contrary to the systems developed in artificial intelligence, which simulate certain capacities of human reasoning in a certain field, based on knowledge structures where a reasoning mechanism of the inference type operates; MAS are conceived and implemented as a set of agents that interact according to modes of cooperation, competition and negotiation, and which therefore continuously conform their behavioral organization to achieve the most effective form each time.

A MAS is characterized by the following:

– Each agent that composes it. It has limited information and problem-solving capabilities. It only has partial or local knowledge and appreciation to the general problem that the MAS must handle and solve.

– The absence of global and centralized control of the MAS, which is the major point.

– The data of the system, which are decentralized and taken by some interface agents of the system, by managing the distribution and the temporality problems.

– The calculation of the solution of the problem that the system must solve at each phase of interpellation. Thus, the functioning of the MAS is achieved by managing the coordination of all the agents in an asynchronous way, in order to release a small set of agents by emergence, which will carry out the action solution of the problem.

Moreover, a MAS can be seen as a set of agents, located in an environment composed of other agents and objects that are different from the agents. Indeed, the agents use objects of the environment (object in the strictly functional and computing sense of the term), which are simply reactive elements providing information and producing functional actions. The agents have a capacity to interpret the information given by the methods of these objects, as well as the behavior of the other agents, taking the time to do so. These agents then use the objects and communicate with other agents, in order to achieve their goals. In the system that we define, this makes it possible to distinguish the set of information to be comprehended – which will be produced systematically by objects whose role it is to do so – and their conceptual analyses and interpretations at multiple levels, which will be the purpose of the organization of agents.

1.6.5. *MAS with reactive agents*

In this kind of system, composed of agents simply considered as reactive, we program a set of reflexive methods to respond to the different events that may occur, by decomposing the actions and carrying them as behavioral actions in agents. The problem to be solved is then the good synchronization of the distributed actions. In each agent, we have what we call a stimulus: an action link to be managed in the right time-frame and according to the state of the environment. The system analyzes the stimuli through its comprehension by the agents that are sensitive to them by nature, finds reflexive methods in certain agents, if they even exist, and then acts through these agents in the most synchronized way possible. These systems can appear to be intelligent if they really work according to what is expected of them, but they do not attach any meaning to their actions, they remain functional. The coordination of the agents does not go beyond the problem of their strict functional regulation, which must be as efficient as possible. In short, these systems were designed to respond to a precise range of events, which made them vulnerable to unforeseen events.

With behaviorally emergent reactive agents, MAS remain among the best-known examples of reactive systems for computing applications in limited and very specific fields.

1.6.6. *MAS with cognitive agents*

These MAS will be able to differentiate information from the external world, by interpreting it through a cognitive symbolization based on various predefined characters housed in the agents' structures. They understand semantic features coming from the information, provided as data to extract the unifying meaning, according to tones. A perceptual system considers a perceived event as a complex fact, and for this purpose, it transforms it into many symbolic characters in relations organized into a group of agents, which contains information on many possible interpretations. Each group of active agents is then a semantic form that will symbolize what is perceived, and the different active semantic forms will perform a multi-scale categorization of the represented facts. It is this type of MAS that we will use. The entire problem will then be the correct consideration of this categorizing form of any phenomenon comprehended by the autonomous system.

In order to design a system that will interpret the current situation of the system in its environment, we will us a massive MAS, where the manipulated entities will all have a proactivity. We define this very important notion:

> An agent is a conceptually based element that is proactive, in other words, active according to its necessity to be so, that uses its knowledge according to its state and that of its context, responding or not responding to requests from other agents.

We use agent organizations for the following two reasons:

– Agents can dynamically reify any specific knowledge in relation to other knowledge represented in other agents. This means that the used knowledge can be considered as aspects in a vast relational organization that will, by dynamic constructions that are continuously made, express the causal and timely relations, as well as the relevant general point of view of the system on its current situation.

– The proactive and very communicative behavior of the agents will allow the formation of aggregations of agents in action and in communication with each other, which can be likened, by transposing it into the social domain, to "social groups". This may indicate, by the evolutionary relations that bring out sets that are more dynamic than others, the existence of a body of knowledge, organized according to a common character that is supported according to a certain light given by the nature of the agents. It is no longer a question of simply solving an optimization problem represented by functions and variables in a certain totally defined space, but of making cognitive forms emerge in the communicative activities of the multiple agents; forms that represent the multiple aspects of the functionalities and decisions that are really adapted to a complicated and evolving situation.

Only agent organizations can satisfy these two characteristics. Objects in object languages are entities that are perfectly adapted to a rational design in predefined structures: all behaviors are foreseen and fully planned. Agents will obviously be implemented with objects and processes, remote objects and threads, with parameter modifications, and the creation of new objects and processes; their conceptual level will always mix activities, knowledge, migration and the creation of new instances or new classes.

2

Reliability of Autonomous Systems

2.1. Introduction

Autonomous systems have the ability to act on their own, in order to carry out actions that are necessary to achieve predefined objectives in response to stimuli which, in robotics, for example, come from sensors. There is the notion of intelligent control and the reliability of the system. Intelligent control involves algorithms, linguistics and mathematics applied to systems and processes (Cardon and Itmi 2018). Let us take the example of the autonomous vehicle, where some autonomous car models are impressive in their reliability and ability to detect obstacles and avoid them. Detection, however, remains too random for the moment: unfavorable conditions (poor visibility, obstacle not recorded in the database) can cause such systems to fail. Current examples of these vehicles include the Tesla X and Tesla S, BMW 5 series, Volvo S90, Audi A8 and Mercedes S class, with BMW offering different levels of autonomy, from the first level offering no assistance to the last level of autonomy, where the vehicle drives by itself. According to several experts, the growth of the Industry 4.0 concept and autonomous systems are the challenges of tomorrow. The expected performances of these systems are the perception of their environment and the ability to interpret different situations by algorithms, based on artificial intelligence or machine learning. Among the systems most used in the industry, there are radars, ultrasonic and inertial vision sensors, and light detection and ranging systems (LIDAR).

Several sectors are affected by the reliability of autonomous systems, such as logistics, robotics, mining, agriculture, autonomous drones, defense, aeronautics, spacial, medicine, etc.

In this chapter, we propose models of the reliability of systems. We provide definitions, the different parameters involved, indicators and the reliability diagram.

2.2. Dependability of a system

The dependability of a system is the set of tools and methods that make it possible to:

– characterize and control the effects of hazards, failures and errors;

– quantify the characteristics of the device or systems, so as to express the conformity in time of their behaviors and their actions;

– analyze the causes of component failure, so as to estimate their consequences on the service provided by the device or system.

Dependability consists of knowing, assessing, predicting, measuring and controlling the failures of systems. This discipline has acquired its name and its current form in different industrial sectors, as a result of its correlation with the notion of quality and ergonomic problems.

Dependability is also called the science of "failures". Other terms exist, depending on the field of application: risk analysis (oil industry), hazard analysis, cyndinics (science of danger) and RMAS (reliability, maintainability, availability, safety). It is characterized by both static and dynamic structural studies of systems, from a predictive and also operational and experimental point of view, taking into account probabilistic aspects and consequences that are induced by technical and human failures.

Dependability is a means or a set of means: steps, methods, tools and a vocabulary. The aim of dependability is to control risks.

The analysis of dependability allows us to place justified confidence in the system under study. This confidence depends on what is given importance, as well as the relative values of these characteristics.

2.2.1. General concepts

The dependability of a system corresponds to its ability to maintain the quality of the service it provides over time.

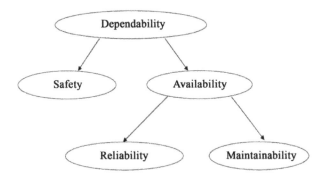

Figure 2.1. *Different components of SoR*

This is a comprehensive concept that is mainly comprised of the characteristics of reliability, maintainability, availability, safety (RMAS) and durability or combinations of these capabilities. In a broad sense, the SoR is considered as the science of failures and malfunctions. It is primarily based on availability and safety (see Figure 2.1).

DEFINITION 2.1.– Safety is the absence of unacceptable damage. It is the probability (see Appendices 1–10) that an entity will avoid critical and catastrophic events under given conditions.

For example, in the computer field, security can mean the following:

– Safety aims to protect against catastrophic failures (e.g. hard disk failure).

– Security corresponds to the prevention of unauthorized access or manipulation of information.

– Security does not necessarily depend on reliability. Indeed, a device can have a low frequency of failure, but when it fails, it can cause failure on another device or on the studied system.

DEFINITION 2.2.– Availability is an entity's ability to perform a required function under given conditions and at a given time. We note this availability as A(t), and we get:

$$A(t) = \Pr[E \text{ not failing at time } t] \tag{2.1}$$

with E being a given event.

To achieve the level of availability, three components are needed: reliability, maintainability and maintenance logistics. Reliability is defined as follows:

DEFINITION 2.3.– Reliability is an entity's ability to perform a required function under given conditions and for a given time. We note this reliability as R(t), and we get:

$$R(t) = \Pr[E \text{ non-failing over } [0,t]] \qquad\qquad [2.2]$$

with E being a given event.

Currently, in a world of competitiveness and performance, it is essential to introduce the effects of reliability of the designed devices.

Indeed, reliability requirements are commonly included in calls for tenders, as well as penalties for non-compliance with these requirements.

The reliability of the manufactured product makes it possible to know its behavior according to its use. After the construction of an experimental design of the defects, several actions can be carried out: the immediate correction of the defect and the consideration of the defect in the design of new products.

In the field of production, knowledge of the reliability of the most failed elements of a production machine is very important. The need for mass production prohibits unplanned production stops due to a failure.

Thanks to a history of failures, the reliability of the element in question can be calculated and a replacement frequency can be deduced. The latter will then be scheduled at the most favorable time for production.

DEFINITION 2.4.– Maintainability is an entity's ability to be restored to a state in which it can perform a given function (the maintenance conditions being predetermined). We note this maintainability as M(t), and we get:

$$M(t) = \Pr[E \text{ is fixed on } [0,t]] \qquad\qquad [2.3]$$

with E being a given event.

Maintainability can be translated as a characteristic that makes it possible to ensure the capacity for maintenance in the best possible conditions. It must be established in the perspective that it is a fundamental element, associated with reliability, for the construction of availability. The application of maintainability can be divided into four stages:

– maintainability allowance;

– analysis of tasks and means;

– possible corrective actions.

Maintainability is also:

– the implementation of failure detection devices that makes it possible to monitor the proper functioning of the studied systems;

– the choice of diagnostic methods.

Maintenance logistics is the policy and means of maintenance. The main maintenance policies are based on:

– corrective maintenance: systematic research for the improvement of the equipment;

– preventive maintenance: slowing down the ageing of equipment, maintaining its level of performance;

– curative maintenance: restoring equipment to working order when it malfunctions.

The reduction of the maintenance intervention time can be achieved by:

– better accessibility of an element in a set;

– ease of disassembly and reassembly of elements requiring frequent interventions;

– interchangeability.

2.2.2. *Failure and repair rates*

Let us begin by defining the notion of failure (Figure 2.2).

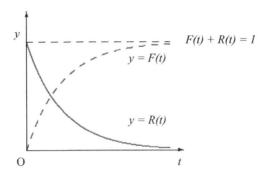

Figure 2.2. *Reliability–failure relationship*

DEFINITION 2.5.– Failure is the end of an entity's ability to perform a required function that passes into the state of failure. We note this failure as F(t), and we get:

$$F'(t) = 1 - R(t)$$ [2.4]

There are different types of failure criteria:

– drift or degradation;

– catalectic: complete and sudden failure;

– total, partial, systematic, major, minor and catastrophic;

– software;

– human.

The reliability approach introduces a quantitative measure of the risk of failure, by means of a probabilistic approach. The quantitative methods used to analyze the reliability of a system are:

– preliminary risk analysis (PRA);

– state space method (SSM);

– fault tree analysis (FTA);

– reliability diagram (RFD);

– failure mode, effects and criticality analysis (FMECA).

We now examine the different indicators of dependability. The failure rate is the limit, between t and t+dt, of the quotient of the failure probability density by the probability of non-failure before t. It is generally noted as $\lambda(t)$, and is written in the following form:

$$\lambda(t) = \frac{1}{R(t)} \cdot \frac{dF(t)}{dt} = -\frac{1}{R(t)} \cdot \frac{dR(t)}{dt} = \frac{f(t)}{R(t)} \qquad [2.5]$$

where $F(t)$ represents the probability of the occurrence of a defect in the interval [0,t], and R(t) represents the probability of success at a given time t.

Let T be the random variable relating to the duration of the entity's operation before failure. We then get:

$$\lim_{t \to 0} F(t) = 0 ; \quad R(t) = \Pr[T > t] ; \quad R(0) = 1 \qquad [2.6]$$

And the distribution function of the random variable T is written as:

$$F(t) = \Pr[T \leq t] ; \quad \lim_{t \to +\infty} F(t) = 1 \qquad [2.7]$$

The probability of failure F(t) for a given period is complementary to the reliability R(t) of the device for the same period considered. We then write:

$$F(t) + R(t) = 1 \qquad [2.8]$$

with $R(0) = 1$ and $\lim_{x \to +\infty} R(t) = 0$.

The failure probability density is:

$$f(t) = \frac{dF(t)}{dt} = -\frac{dR(t)}{dt} \qquad [2.9]$$

and f(t) dt is the probability of failure of the entity between t and t+dt:

$$f(t)dt = \Pr[t < T \leq t + dt] \qquad [2.10]$$

from which the distribution function F(t) is written in the following form:

$$F(t) = \int_{-\infty}^{t} f(t) \cdot dt \qquad [2.11]$$

$\lambda(t)$ dt is the probability of failure over [t,t+dt], knowing that the entity has not failed over [0,t]. Hence:

$$\lambda(t)dt = \Pr[t < T \leq t + dt \mid T > t] \qquad [2.12]$$

So:

$$\lambda(t)dt = \frac{\Pr[(t < T \leq t + dt) \cap (T > t)]}{\Pr[T > t]} \qquad [2.13]$$

As $(T > t) \cap (t < T \leq t + dt) = (t < T \leq t + dt)$.

$$\lambda(t)dt = \frac{\Pr[(t < T \leq t + dt)]}{\Pr[T > t]} \qquad [2.14]$$

Hence:

$$\begin{aligned}\lambda(t)dt &= \frac{f(t)dt}{R(t)} \\ &= -\frac{dR(t)}{R(t)}\end{aligned} \qquad [2.15]$$

where $\int_0^t \lambda(x)dx = \int_0^t -\frac{dR(x)}{R(x)} = -\ln R(t)$ and $t = 0$, we get $R(0) = 1$. The reliability is therefore written in the following form:

$$R(t) = e^{-\int_0^t \lambda(x)dx} \qquad [2.16]$$

In the special case where the failure rate is constant, we get:

$$R(t) = e^{-\lambda t} \text{ and } f(t) = -\frac{dR(t)}{dt} = \lambda e^{-\lambda t} \qquad [2.17]$$

2.2.3. Average estimators

Currently, we use the average times of operation of the systems. Let us start by defining the MTTF.

DEFINITION 2.6.– The MTTF (Mean Time To Failure) is the average of the operating times from time 0 to the first failure.

This time can be calculated using life tests. The MTTF is generally defined as follows:

$$MTTF = \int_0^{+\infty} t.f(t)dt$$

$$= -\int_0^{+\infty} t \cdot \frac{dR(t)}{dt} dt \qquad [2.18]$$

$$= [t.R(t)]_0^{+\infty} + \int_0^{+\infty} R(t)dt$$

At $t = 0$, $R(t) = 1$, where $t.R(t) = 0$. As t increases, R(t) decreases, so k exists such that $R(t) < e^{-kt}$ as:

$$\lim_{t \to +\infty} t \cdot e^{-kt} = 0 \Rightarrow \lim_{t \to +\infty} t.R(t) = 0 \qquad [2.19]$$

So: $MTTF^* = \int_0^{+\infty} R(t).dt$.

DEFINITION 2.7.– The MTBF (Mean Time Between Failure) is the average time between two successive failures. It is only defined for repairable systems.

The MTBF can be determined by testing the system over a period of time T and counting the N failures that have occurred:

$$MTBF = m = \frac{T}{N} \qquad [2.20]$$

In a special case where λ is a constant, we get:

$$MTTF = \int_0^{+\infty} e^{-\lambda t} dt$$

$$= \left[-\frac{1}{\lambda} \cdot e^{-\lambda t} \right] \qquad [2.21]$$

$$= \frac{1}{\lambda}\bigg|_0^{+\infty}$$

Consequently:

$$R(\text{MTBF}) = e^{-\lambda MTBF}$$

$$= e^{-\frac{MTTDP}{MTMP}} = e^{-1} = 0.37 \qquad [2.22]$$

This means that after a time t=MTBF, approximately 63% of the components that are in working condition at the beginning of the test will fail.

The MTBF is only equivalent to the MTTF for a population of entities with a constant failure rate. The MTTF is defined for non-repairable systems and the MTBF for repairable systems.

DEFINITION 2.8.– The mean up time (MUT) is the average time of availability after repair.

This average time is slightly lower than the MTTF because, in this case, the repair may be partial, especially in the case of redundant structures, whereas the MTTF characterizes the average operating time of a completely restored entity.

DEFINITION 2.9.– The MTTR (Mean Time to Repair) is the average of the technical repair times. By definition, $\text{MTTR} = \int_0^{+\infty} t.m(t)dt$, which gives:

$$MTTR = \int_0^{+\infty} [1\text{-}M(t)] \, d\,t \qquad [2.23]$$

DEFINITION 2.10.– The MDT (Mean Down Time) is the average time of unavailability after repair.

This time includes: the technical time to detect, locate and repair the failure, and also the administrative time for management and transport. The MDT is higher than the MTTR, but generally lower than the MUT. The following relationship exists between estimators:

$$MTBF = MUT + MDT = \frac{1}{\lambda} \qquad [2.24]$$

The last equality is true if λ is constant. Figure 2.3 represents the indicators of predictive reliability.

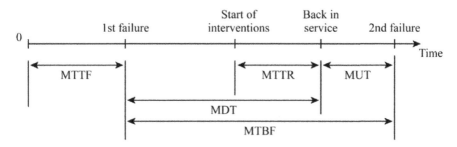

Figure 2.3. *Different indicators*

2.2.4. *Some methodological tools*

To give an idea of the terms and tools used in FMD and to present the method for assessing the predicted availability of a means before its realization, it is necessary to follow the frequency of occurrence of stoppages (reliability field) and to know the repair times (maintainability field) of the means, and thus of each of its sub-sets. In order to achieve this task, we will only present two methods: the FMECA method and fault tree analysis (FTA).

2.2.4.1. *The FMECA method*

The FMECA method is the most cited and used in companies. Its principle is as follows:

– decompose the system into components. The decomposition is determined on the research of failure modes and frequencies;

– associate its failure modes to each component by listing all the failure modes;

– identify the effects on the system of each failure mode of each component;

– define the criticality scale;

– determine the effects and consequences of the failure modes of the components;

– determine the system-level consequences of mission success.

The purpose of this cause/consequence analysis is to draw conclusions. The various official standards on FMECA thus include suites relating to the exploitation of this analysis, which are somewhat different depending on the framework in which the method is used.

2.2.4.2. Fault tree

The representations of logical links between causes and consequences that start from an event have a tree-like form. The most commonly used are: cause tree, event tree and fault tree.

In this section, only the fault tree will be discussed. The fault tree is the representation of various possible combinations of events that can lead to a single undesirable event. It uses all the logical "and" and "or" connectors, both inclusively and exclusively, as well as their combinations.

The construction of a fault tree consists of selecting an event whose failure scenarios we want to represent. The answers found are represented with logical connectors ("or", "and" possibly exclusive "or").

The reliability analysis of a system can be achieved using the fault tree method in four steps:

– review of the system and identification of potential adverse events in the system;

– modeling of the system, research and description of the events that can occur during its life;

– establishment of fault trees;

– calculation of the probability of undesirable events and reliability.

2.3. Reliability diagram

The reliability block diagram (RBD) method was the first technique used to analyze reliability systems. The cause tree method (CTM) and the failure mode and effects analysis (FMEA) came later. It is currently used in many industrial sectors for non-repairable systems. It can also be used under certain conditions for repairable systems. A reliability diagram describes the logical links between the components that are essential to the success of the system's mission.

This description is realized by a graphical representation of the functional behavior of a system in the form of blocks. The reliability diagram method is also called success path. Generally, a success path is defined as a path from the far left to the far right of the diagram.

The purpose of reliability diagram analysis is to represent the operation of a system. Block modeling represents components or sub-systems. This modeling consists of looking for the links between the sub-systems. Reliability diagrams are logical functional diagrams that translate the consequences of the behavior of each element on the global behavior of the system. We present the principle of this modeling by series, parallel or mixed systems.

2.3.1. *Series system*

DEFINITION 2.11.– A system composed of several sub-systems is described to be in series if the failure of one of its components leads to the failure of the entire system.

The series system is represented by Figure 2.4. A system composed of at least two sub-systems of event E_i, with the failure of one of them leading to the failure of the system "event E":

$$R(t) \quad = \Pr[\bar{E}] \quad = \Pr\left[\overline{E_1 \text{ or } E_2 \text{ or } \dots \text{or } E_n}\right]$$
$$= P\left[\bar{E}_1 \text{ and } E_2 \text{ and } \dots E_n\right] \tag{2.25}$$

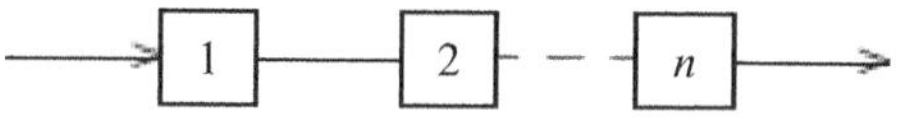

Figure 2.4. *Series system*

Given that the sub-systems are independent in terms of their failures, we get:

$$R(t) = \prod_{i=1}^{n} R_i(t) \tag{2.26}$$

where $R(t)$ is the reliability of sub-system i.

2.3.2. *Parallel system*

DEFINITION 2.12.– The system is defined as in parallel (Figure 2.5) if it works as soon as at least one of the sub-systems works, its failure implies the failure of all the sub-systems.

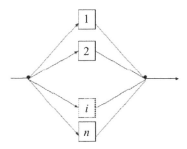

Figure 2.5. *Parallel system*

The failure $F(t)$ of a parallel system is written in the following form:

$$F(t) = 1 - R(t) = \Pr\left[E_1 \ and \ E_2 \ and \dots and \ E_n\right] \tag{2.27}$$

If the sub-systems are mutually independent, then the system reliability is written as:

$$R(t) = 1 - \prod_{i=1}^{n}\left(1 - R_i(t)\right) \tag{2.28}$$

In the case of a constant failure rate, we get $R(t) = 1 - \prod_{i=1}^{n}\left(1 - e^{\lambda_i t}\right)$ and:

$$\text{MTTF} = \sum_{i=1}^{n}\frac{1}{\lambda_i} - \sum_{i}\sum_{j\neq i}\frac{1}{\lambda_i + \lambda_j} + \sum_{i}\sum_{j\neq i}\sum_{k\neq j\neq i}\frac{1}{\lambda_i + \lambda_j + \lambda_k} + \dots + (-1)^{n+1}\frac{1}{\sum_{i}\lambda_i}$$

In the case where all elements have the same λ, we get:

$$\text{MTTF} = \frac{1}{\lambda}\sum_{i=1}^{n}\frac{1}{i} \tag{2.29}$$

2.3.3. *Mixed system*

2.3.3.1. *Parallel–series system*

A mixed (parallel–series) system is shown in Figure 2.6.

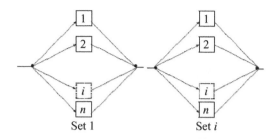

Figure 2.6. *Parallel–series system*

The reliability of a set j is given by:

$$R_j(t) = 1 - \prod_{i=1}^{m_j}\left(1 - R_{ij}(t)\right)$$
[2.30]

where $R_{ij}(t)$ is the reliability of element i of the jth set. Therefore, the reliability of the entire system is written as:

$$R(t) = \prod_{j=1}^{n}\left[1 - \prod_{i=1}^{m_j}\left(1 - R_{ij}(t)\right)\right]$$
[2.31]

2.3.3.2. *Series–parallel system*

A mixed (series–parallel) system is shown in Figure 2.7.

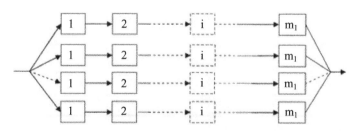

Figure 2.7. *Series–parallel system*

The reliability of a series branch is written as:

$$R(t) = \prod_{i=1}^{m_j} R_{ij}(t)$$ [2.32]

The reliability of the set is written as:

$$R(t) = 1 - \prod_{j=1}^{n}\left[1 - \prod_{i=1}^{m_j} R_{ij}(t) \right]$$ [2.33]

2.3.4. *More complex systems*

For these systems, we use the total probability theorem (see Appendices 1–10 on probability).

EXAMPLE 2.1.– Let us assume that we have a system with shared redundancy. In the aa' and bb' paths, it has been found that a and b have insufficient reliability. We then added the component c in shared redundancy: it makes it possible to add the ca' or cb' success paths.

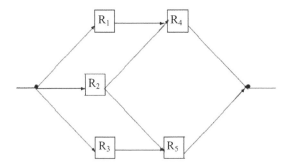

Figure 2.8. *Complex system*

Consequently, we get:

$$1 - R_s(t) = \left[1 - R_{a'}(t) \right]\left[1 - R_{b'}(t) \right]R_c(t) + \left[1 - R_a(t)R_{a'}(t) \right]$$
$$\left[1 - R_b(t)R_{b'}(t) \right]\left[1 - R_c(t) \right]$$ [2.34]

2.3.5. *Fault tree*

A fault tree is a logic diagram using a tree-like structure, linking the failure of elements to a particular failure state of the system. The undesirable event should be defined as precisely as possible. The fault tree graphically represents the combinations of events that lead to the undesirable event at the top of the tree. It is made up of successive levels, such that each event is generated by the lower levels through various logical operators. The fault tree, like the reliability diagram, is the representation of Boolean relationships between elementary events. It can be solved by Boolean algebra. There is a correlation between the reliability diagram and the fault tree, thus:

– a series diagram is an "or" gate;

– a parallel diagram is an "and" gate.

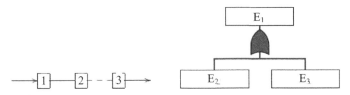

Figure 2.9. *Series diagram*

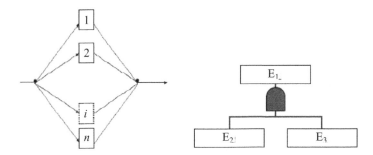

Figure 2.10. *Parallel diagram*

Reliability diagram and fault tree modeling are similar, but fault trees can be built step-by-step, which is not the case for reliability diagrams.

2.4. Reliability networks

DEFINITION 2.13.– A reliability network is a representation of logical operating or malfunctioning conditions by means of an input and output dipole.

This representation is particularly convenient for defining failure conditions. We define a reliability network R on a set $e=\{e_1, e_2, ..., e_n\}$ by:

1) the r-applied graph G = (S, U, α, β) without loops, in which two vertices of (O) origin and (Z) extremity exist, where r is the maximum number of arcs with the same extremities;

2) the Δ application of U in e, such that:

$$\forall u_i, u_j \in U, \alpha(u_i) = \alpha(u_j) = S_n$$

and $\beta(u_i) = \beta(u_j) = S_m \Rightarrow \Delta(u_i) \neq \Delta(u_j)$.

We denote it as: R=(S,U,α,β,Δ), with S being the set of vertices; U being the set of arcs; α and β are applications that give the vertices of the extremities of an arc; and Δ is the application that maps each arc of the graph to a component of e.

2.4.1. *Partial graph associated with a subset of components*

By definition, from an R = (S, U, α, β) and e' included in e is a subset of components. The partial graph associated with e' is: Gp(e') = (S, Up(e'), α, β) such that: Up(e') = {u ∈ U / Δ(u) ∈ e'}. In other words, we only keep arcs in U that are associated by Δ with components of e'.

DEFINITION 2.14.– We call a reliability R network link a subset a \subset e of components, such that there exists a path from O to Z in the graph Gp(a).

Any link corresponds to one or more paths of the graph. A path u = (u_1, u_2, ..., u_n) from O to Z corresponds to the link formed by the images of $u_1, u_2, ..., u_n$ by Δ.

DEFINITION 2.15.– A network of reliability R is a subset of components b of e, such that the subset of arcs Up(b) contains a cut of the graph G. A section of R can contain several cuts of G.

2.4.2. *Reliability network and structure function*

Let e_1 be the subset of components in good condition (respectively failing components). The system works if e_1 is a network link of reliability R, (respectively failing in the opposite case). If the system works, its structure function ϕ $(x_1 \ldots x_r)$ that is 1 (respectively ϕ $(x_1 \ldots x_r)$ is 0. The reliability network and the structure function of S have the same cuts:

– link a is minimal if no subset of a is also a link;

– section b is minimal if no subset of b is also a cut.

Additional cut/link:

– if a is a link, then $\overline{a}$ is not a cut;

– if b is a cut, then $\overline{b}$ is not a link;

– a network is known as a degenerate network if it has no link or cut.

2.4.3. *Properties of reliability networks*

A number of properties set out for structure functions can thus be available for reliability networks, using the following properties:

– In a reliability network, a subset of components including a link and a link, a subset of components including a cut and a cut.

– A cut includes at least one component of each link and a link includes at least one component of each cut.

– Any link includes at least one minimal link, any cut includes at least one minimal cut.

– In a non-degenerate network, a set b of components and a cut if – and only if – it contains at least one component of each minimal link.

– In a non-degenerate network, a set a of components and a link – if and only if – it contains at least one component of each minimal cut.

2.4.4. *Length and width of a reliability network*

Theorem: the number r of components of a reliability network is at least equal to the product of the length and the width of the network (El Hami and Radi 2013b):

$$r \geq L.l \quad \text{(Shanon 1948 and Moore 1965)}$$

with L being the network length and l being the network width.

Typical networks are as follows:

– series network: each component is a cut, the set of components is the only link;

– parallel network: each component is a link, the set of components is the only cut.

2.4.5. *Equivalence between structure function and reliability network*

A component X_i of a reliability network is useless if any link containing X_i is still a link after X_i is deleted (any cut containing X_i remains a cut after X_i is deleted).

2.4.6. *Construction and simplification of reliability networks*

If we know all the minimal links, we obtain an equivalent network by placing k sub-networks in parallel, consisting of each of the components of a link placed in series.

Similarly, if we know all the cuts, we obtain an equivalent network by placing k sub-networks in series, formed by paralleling the components of each cut. We know that their knowledge allows us to easily find a network of reliability R and/or an equivalent structure function ϕ. It is therefore necessary to determine them.

Let C be the set of elementary paths, with initial extremity O and terminal extremity Z of the graph G of a reliability network. With any elementary path $\mu= \{u\_1, ..., u_j, ..., u_e\}$, we associate the link a(μ) by Δ, formed by the components corresponding to the arcs of the path. Δ is the application associated with the r-applied graph defining the reliability network. It associates a component of the system to each arc of the graph:

$$a(\mu) = \{\Delta(u_1), \ldots, \Delta(u_i), \ldots, \Delta(u_e)\}$$

Let L be the set of links obtained from C:

$$\mathcal{L} = \cup_{\mu \in C} \{a(\mu)\}$$

The set L then includes all minimal links of the network.

To determine the set C of elementary paths in a practical way:

– for a simple network, it is quite easy;

– for a complex network, we will use the methods for finding paths in a graph. Generally, two methods are used to find the elementary paths in a reliability network: the tree structure and the Latin decomposition.

As far as the tree structure is concerned, we go from the starting point and look for the vertices that can be reached by an arc, and for each of them, we proceed in the same way until we reach the arrival vertex.

DEFINITION 2.16.– A tree with root A is called a graph without a circuit, such that:

– A is a vertex which is not the terminal end of any arc;

– any vertex other than A is the terminal end of just one arc.

It is therefore a 1-application graph. To represent it, we divide the vertices into levels. The length of a path from A to a vertex is equal to the level of the latter.

For the Latin composition method, we are interested in all the paths from O to Z; the paths leading to O from Z are insignificant. Moreover, these are multi-graphs (r-applied), so we have to reduce to a simple graph (1-applied) and in the list of obtained paths, multiply the multiple arcs.

3

Artificial Intelligence, Communication Systems and Artificial Consciousness

---

3.1. Introduction

Today, computer science is a central discipline in science and in society because of the innumerable uses of software that constantly communicate. We will detail its history, show how artificial intelligence has become significant and demonstrate that we are heading towards a distributed, highly communicating and autonomous artificial consciousness.

These continuous communications between humans through computerized systems come from a tendency to communicate, which human technology has very highly developed, and whose source comes from the informational substratum of the Universe, which is essentially communicational.

3.2. Evolution of computer science

The first computers were created in the middle of the 20th century: the Mark I dates from 1943 and performed 10 instructions per second, and the ENIAC was created in 1946 and was able to process 300 instructions per second. These totally local machines used lamps in their electronic circuits, and processed small sequences of instructions in order to perform numerical calculations. Today's most powerful computers use electronic components and compile billions of instructions per second.

Computer science can be seen in three different aspects, which are, in fact, complementary:

– It is the science of calculable models. This science, which is abstract and initially close to the numerical models of mathematics, studies everything that can be conceived – and then effectively calculated – using elementary instruction sequences, representing the stages of function calculations with algorithms and programs.

– It is computer and application software technology, carrying out processing and calculation tasks applied to an innumerable number of fields. This technology is situated within the framework of the activity of abstract objects, which have been conceived and manufactured according to processes that are specific to companies and researchers of all disciplines.

– It is the market of all computer realizations, in the form of remotely accessed computers and software, with means of communication and access by networks to servers. It is the very important trade of these products: the field of supply and demand.

Computer science is therefore both an abstract and formal science, based on the notion of modeling what can and will be calculated, as well as cutting-edge technology that is very closely linked to the development of systems and devices in all companies. This represents a considerable market. Many companies, even some scientists, consider computer science as a technology that is necessarily applicable to their fields, being oriented towards the action of effective achievements.

Nowadays, a computer is the medium by which many programs can run concurrently and which can also, in some cases, modify and rewrite themselves. This ownership of autonomous rewriting of programs was already used a long time ago: in 1960, with the LISP language, which, at that time, provided the conditions for rewriting in programming. What computer science achieves and allows today can be measured in the processing of programs, which produce the values of innumerable functions, as well as make predictions in a great many fields that are essential to production and the economy in all societies. However, computer science is not limited to these productions that grow and allow the development of the economy; it is not just a sophisticated technology that puts predefined programs into action, which run on computers in networks.

Generally speaking, computer science deals with the processing of information, which is related to sequential calculations of functions by systems using state machines as a basic element. A state machine is an abstract machine passing through states that are strictly determined by the program, choosing them from an available set, and in which elementary instructions are executed. We use the state machine by starting from an initial state, then reaching a final state which will be the result of the calculations. This notion is fully used when dealing with problems that have been decomposed into many sub-problems, but which are all well-defined, with the whole forming a perfectly established structure where what is to be calculated is well known, allowing the development of the precise sequence of what is to be calculated by the successive operations.

Such problems belong to the class of well-decomposable problems. At the elementary level, computer science deals with binary information encoding elementary instructions, the instructions forming the programs which will thus be sequences of calculations. The length of the programs and their number of instructions can be considerable, and several programs can run simultaneously and communicate information to each other at the right time. It is therefore common that the programs are designed with many sub-programs, which treat specific sub-problems and which, together, form a structured program that is made to treat all the elements of the problem.

The mechanistic vision of computer calculations has evolved. Today, we know how to make many programs communicate, based on state machines, which run in parallel and, above all, which modify their state machines themselves during or after their operation, even though the basis of each program always remains a state machine. We have thus moved the framework of program automation to the notion of autonomy. We know how to build programs made of many sub-programs which have their own behaviors, which can communicate at the right moments by calling to one another, synchronizing themselves, modifying themselves, thus generating new programs, breaking the order initially given by the state machine defined by the programmers. These are multi-process and self-modifying systems, and it is these systems that currently run on all computers. This is, indeed, the case for any operating system, as well as any website that manages resources and simultaneously active applications for users. The notion of the autonomous process is important, and it will be the same for artificial consciousness, because it places the consideration of programs at the same level as active autonomous software entities, carrying out precise local actions, but above all communicating with one another, forming dynamic organizations that are constantly modified by their very functioning. These processes will be generalized with the notion of software agents, which has been highly developed in artificial intelligence. Software agents are autonomous programs formed by process structures; they are very communicative and can modify themselves according to their functioning, as well as create new agents.

Broadly speaking, there are therefore two main categories of programs:

– The category of programs where it is a question of calculating a precise and very well-defined function, of strictly developing the calculations of all the stages, which then come back to the execution of a very structured set of state machines.

– The category of autonomous programs composed of multiple specific sub-programs with a high degree of autonomy, some of which run in parallel, which will seize information that is external to them, that is foreseen or unforeseen for them, making them confront each other at certain moments to exchange information, which will modify them and generate other sub-programs that are highly adapted to the new situation, in order to produce a global result that will be as adapted as possible to the framework of use that has evolved.

The first category is, for example, that of a user of software to help build a large mechanical system, who enters parameters about the system they have to build by selecting them on a graphical interface, and then obtains the manual characters to be made in the right order by using references of the mechanical system, housed in databases. The second category is, for example, the state of Internet users' programs over a specific period of time, when these programs themselves are constantly viewing and modifying highly interactive websites. There are no permanently fixed elements in this second case, and the problem cannot be based on an a priori decomposition into independent elements. The difference between these two categories of problems is based on a fundamental point: there are programs in both cases, but in the second case, they will have to be modified, rewritten and evolve, whereas in the first case, the software used to help build a technical element is used as it was designed, with its initial capabilities being well defined and considered permanent.

Computer science as a science of computability appears on integer mathematical functions. What mathematicians can define on the integers in the form of functions and multiple equations, are equivalent to computer programs. It has been shown that for any function, based on a sequence of integers and producing another sequence of integers, to make mathematical sense, there must be some abstract machine, some abstract computer with instructions to calculate it. The existential of all mathematical functions on integers has a meaning if computability allows it to have one, and vice versa. This theoretical result is Church's famous thesis from 1936. It amounts to saying that for a function on the integers to have a mathematical meaning, the program of a theoretical machine which can compute it must be defined.

The use of computer programs has evolved towards programs based on the representation of knowledge of specific fields and endowed with autonomy, being able to constantly improve themselves by analyzing the data they receive and the results they provide, which is the framework of artificial intelligence. Therefore, in the realization of these programs, there are the following three steps:

– The very fine representation of the conceptual field to be constituted, with the development of all the necessary cognitive and functional structures.

– The coding of the conceptual model, with the representation of the functionalities to be calculated in the form of multiple sub-programs and the precision of the improvements that the system must itself bring to its structure, by assessing its productions.

– The functioning of the program by using it with numerous data, in order to refine its structure and to make it truly operational.

3.3. Evolution of artificial intelligence

Artificial intelligence is, in the field of computation, the study and programming of the mechanisms of reasoning and use of knowledge in all fields. It validates its theories and applies them by systematically developing programs. It is therefore an application of the general problem of the representation and use of knowledge, within the framework of computer processing, which represents this knowledge.

AI, like many scientific theories today, programs its models by developing systems, which makes it possible to validate or invalidate the discovered and proposed models, to see whether or not they correspond to reality and are consistent with the uses of knowledge. The question raised and treated is that of the adequacy between the models developed and the relevance of what is produced.

The search process used in AI to address a knowledge problem has the following form:

– definition of a certain level of formal determinism in the processes considered, in order to address the problem;

– elaboration of a decomposed general model into multiple sub-models, which are all programmable;

– experimentation by development and validation or rewriting of the programs: the validation is achieved by comparing the results with the human assessments of the specialists.

AI is thus a discipline included in computer science and which has an applicative aspect, allowing it to build many particular and usable systems, so as to produce reasoning and manipulate knowledge.

Artificial intelligence was officially born in 1956, during a famous conference where the term itself was defined: "The Dartmouth Summer Research Project on Artificial Intelligence". The term "artificial intelligence" is understood in the following way: it is, in computer science, a question of reproducing our capacity to understand, to identify, to predict, to explain and to learn in situations that are often difficult, and where experience and reasoning are preponderant. It was not a question of producing artificial systems with autonomous consciousness, but of finding out how a machine with a computer could become endowed with forms of consciousness, linked to particular reasoning (Newman and Jefferson 2004). These questions were asked at a time when it was not clear how human consciousness worked.

Very powerful results have been obtained in specific cognitive fields where the structures of reasoning were very well assessed, such as medicine, fault diagnosis, games, comprehension of a text in a particular field, image recognition, vehicle piloting and robot behavior, but not yet in the specific manner in which human thought is generated on innumerable subjects that are constantly linked and developed, through using its organized memory and its tendencies.

We therefore consider a problem which is functionally and clearly put forward in its field, with a precise framework where certain applicative facts are known, and we seek its solution in terms of causes and consequences by relying on a vast body of knowledge and data, linking local causes to their consequences. This knowledge, produced and validated by experts in the various fields concerned, is available and is a matter of finding the right knowledge to be used, in order to construct the conceptual structure producing the results, starting from certain facts put forward as a deductive hypothesis. To facilitate this research, knowledge must be highly structured, and much attention has been paid to the discovery of good ways of structuring this knowledge. Such a search therefore amounts to considering a large number of factual relations between cognitive elements, which together form a decomposable whole, in order to extract the most relevant relations at the right moment, and which will generate the continuation of rational reasoning leading to the solution. It is therefore a question of constructing sequences of elementary reasoning based on the notion of chain of causes and consequences, which is called the sequence of the rules of inference.

These research works have focused on the most efficient treatments possible, applying AI techniques to all fields and trying to find powerful models of knowledge representation. However, these research works have favored the combinatorial side in the treatment of problems, with the prior arrangement of masses of data, and have not considered the aspect of intentional generation of knowledge in the ideal representations. This has led to the resolution of many problems, but has not made it possible to solve the deep problem of the motivation of understanding things perceived by an artificial system hosted by a corporeality, which is part of another field taking into account the intentional generation of ideal representations.

This research has not really taken into consideration two aspects of human thought:

– On the one hand, common sense, with the use of common intuition as a means of linking dissimilar objects which are not linked in any causal way, but which have analogies of forms and sensible comprehensions as to their representations, because some of them have already been observed.

– On the other hand, the fact that most reasonings are not situated at a single level of conceptual development, but simultaneously mix different levels of conceptualization and sensible comprehension, with multiple feedback loops linking and conforming these levels.

In fact, in AI, we are not interested in the functioning of the human mind dealing with usual and sensitive problems, which always refer to emotion and to basic tendencies or impulses through sensitive and emotional memorization, but instead, in the transposition of rational reasoning, as sophisticated as possible, we are interested in applied computation. However, these strictly rational reasonings do not include the notion of affectivity, emotion, pleasure and displeasure, nor the notions of curiosity and inhibition, which ultimately lead to the creation and generation of original concepts, and which have created human society by opening up to ethics.

There is yet another field that has not been well considered by AI, but which is beginning to emerge today in DAI (distributed artificial intelligence). It is the notion of knowledge sharing in social organizations. In particular, any emergence of a solution to a problem of reasoning that is not reduced to being linear by a local structure is the result of multiple parallel developments of several organizations that have been in confrontation and in co-activity at a certain moment and at a certain level, and whose expression of the solution is at another level. The process of generating thought is distant from the linear trace of formulations that has been put simply in the form of symbolized words, with the search for the right fact to be found in a very large set where it would already be found. To think in life is very rarely to develop a chain of relations that make it possible to select good sequences with causality, except in some specific works. An ideal generation is very specifically a totally constructed organizational emergence, expressing a posture of the thinking being, which takes place at another level than that of the elements that have allowed this emergence. This is the thesis that we have strongly developed in our work on artificial consciousness (Cardon 2018).

What has not been considered by AI is, in fact, the very deep complexity and intention of the productions of ideas leading to intelligent productions. Thinking is not reduced to the manipulation of symbols, it is not the unfolding of a program. Initially, AI did not want to consider the reason that could engage a system to operate factual generations, like the way the mind generates thoughts with sensible reasons to do so, and reasons that are only rarely decisions to solve rational problems. Systems that are traditionally developed in AI have neither desire to function nor pleasure in functioning, and feel nothing in their efforts to solve problems. These characteristics of appreciation, which are typical of the human mind in its basic functions, have been considered as outside the field of research because, on the one hand, it was not clear how to approach them and, on the other hand, because the technological demand for applications was preponderant over fundamental multi-disciplinary research including psychiatry and linguistics. However, an unquestionable fact remains: AI is one of the compulsory paths to reaching the conception of artificial consciousness, and most of its results will be used, by placing them locally in the right frameworks of expressions.

Artificial intelligence is a discipline that was created in computer science, but that has taken on a lot of autonomy and that has a multi-disciplinary nature. In the themes of its research and development, it associates computer scientists specializing in many fields, such as the coding and processing of software agent systems and neural networks, as well as researchers in the cognitive sciences, mathematicians specializing in digital technology, automaticians specializing in robotics, sociologists, psychologists, physicians, linguists, etc. Its developments tend to very finely imitate the cognitive activities of analysis, deduction, reasoning and often unify them if they are in different fields. Moreover, it is currently developing, in the behavior of robots, the representation of human emotional characters, as well as tendencies to think about certain things and have certain desires. This is the path to the maximum autonomy of systems.

3.4. Radical evolution of computing and AI towards fully communicating systems

The systems and software used on computers have been continuously developed, and one of the highlights is the development of autonomous means of communication between the systems. The days of the computer centralizing information sent by a few employees in the rooms of a work place are totally over. This major innovation is the ability to communicate strongly between personnel and to exchange information with other remote systems: knowledge-type information or questions in the form of sentences or images – which are automatically put into digital form so that they can be interpreted by remote processors – deployed on these remote systems and processed, so that they can give rise to the requests and actions that they represent. Communications use wireless networks or fiber optics to transfer phenomenal amounts of digitized data at the speed of light, between any point in a country or continent. With a laptop or a smartphone, we can communicate with anyone from virtually anywhere using wireless networks.

This technology is not without danger, because electromagnetic waves are not neutral on the functioning of the human neural network and on the genomes, but this is another matter where the precautionary principle is not applied.

Receiving and transmitting antennas that are 100 meters high allow communication over a distance of 100 kilometers; cities are saturated with small antennas, allowing multiple wireless communications. These wireless means make it possible to transmit and receive informational flows between innumerable transmitters and receivers, representing texts, images or films.

Nowadays, a laptop and a modern smartphone are a means to communicate digital information, to exchange pictures or e-mails, to query any website, to buy and sell on specific commercial sites or to watch TV shows on countless channels. These devices allow us to take digital pictures that are automatically dated to the second they were taken, and located at the GSP location where they were taken too. They are very easily transferred to personal computers, constituting an important event memory for the user. However, do these memories remain confidential to the user? Google's Android system, used for mobiles, also keeps track of all movements. This makes it possible to create software that informs users that one of their friends is in a certain place near them, and that they can go and meet them: the smartphone warns this friend of the possible meeting. The global memorization of all these local memorizations of trips does not pose any problem for the current large storage systems. This geolocation problem is the same for all smartphone owners, and these locations are commonly stored in the memories of these smartphones in an encrypted way, so that access is restricted. The transceiver centers, of course, have access to it since they know the encryption keys they have set up. It is possible to think that, by not doing anything wrong, you can track and remember all your journeys, but it is less obvious than that. When certain services want to know the specific profile of a person, and when that person is given responsibilities, it is then very useful to know their meetings, their network of relations, their profile defined by their movements, the synopsis of their emails on the Internet and their discussions via smartphones, their habits – including things bought by credit card – which are also memorized piece by piece. All communication actions of all people on the Internet are stored in Europe for a certain period of time, ranging from a few months to two years depending on the European country, the reason given being the possible search for evidence in the case of crimes.

The new DTT televisions will have processors and will be able to connect directly to the Internet to download films or documents, obviously paying for the downloads, with secondary screens where each member of the family will be able to watch what they want interactively. The distinction between a desktop computer and a television will disappear, creating homes where everything will be communicating, that is, where all the electronic devices will communicate by request or even automatically, and where the link with the networks will be constant and uninterruptable under penalty of being "cut off from the world". Does this respect the intimacy of the user or does it not tend to make them a consumer who is continuously assisted and immersed in a virtual world, where the relationship with the other is no longer real?

Internet is a formidable technical creation with undeniable social characteristics. It is a network of networks on a planetary scale that allows computers that have the means to identify themselves, thanks to a particular communication protocol known as the Internet Protocol, or IP protocol, to communicate with each other. This communication is achieved through access providers, on which the user is hosted by their own wire or wireless connection, which recognizes them by identifying them and which will establish contact between the recipient and the sender. Then, thanks to particular software installed on the user's site (called browsers), this connection offers the means of communicating with all the services available on the many machines accessible via the Internet network. These available services are accessible page by page, from downloading files containing documents in different formats such as images, movies and obviously to commercial transactions through messaging. However, there are two consequences to this opening on the global network:

– By communicating with the network, the user declares themselves and therefore identifies themselves. They are identified by their computer number, which is in the components and which is unique, and their activities can thus be followed. The user has many protections limiting intrusions and guaranteeing the confidentiality of exchanges, but all these protections remain and will always remain liable to fail.

– By communicating request information or sending messages, the user sends information on this network, and before arriving at the recipient, this information will be routed on certain nodes of the Internet network, that is, on certain computer systems used to route the packets of information from each sender to each recipient, which is called routing. These packets of information are thus transported from node to node from source to destination, and it seems obvious that, in some cases, some particular services of some States wish to access the nodes in order to learn about the information exchanged. To do this, some users choose to encrypt their information, which is the case for the successful completion of commercial and financial transactions that require the handling of confidential information. Visiting a site puts the user who accesses it under a certain dependence on the site manager. These sites are obviously capable of recognizing the address of the machine that visits them, and they can modify themselves thanks to powerful software automata, under planned aspects, in order to personalize themselves for each identified visitor who has a specific profile, thus engaging their buyer behavior.

The Internet is a wonderful means of communication, linking all computer users and making this network indispensable. It makes it possible to spread information, to create websites with important information and to oppose the restriction of information.

However, no computer system can be totally secure when it is made to be accessible by users that are not referred to by very special filtering protocols, such as those used in very secure places. It can also be noted that the technological race to place players or online traders in stressful situations leads to the paradoxical situation where the player who is led to generate stressful situations through the game, causing them to lose their control, will be offered special bracelets that indicate their stress levels and send messages on a window of their screen to allow them to leave the game.

The Internet is a system that is highly evolutionary and unstable. There is no system composed of a large number of proactive elements, in other words, elements that are active on their own behalf, which can form a stable organization over time. The living, which knows how to stabilize itself at certain scales, is not an accumulation of independent and uncoordinated autonomous entities. There is a multi-scale level of control, and this level is the key to organized life.

The Internet could become a major surveillance system. This surveillance has its limits today; it is not totally efficient because there are too many things to monitor. The intercepted information must be analyzed, to be placed at the semantic level of knowledge, using complex rational systems and human operators who decide on the relevance of certain knowledge.

These surveillance systems do not know how to fully analyze, understand, synthetize and use the information that they capture independently, because they are functional, but research in AI tends to make them perfectly efficient in the analysis of all accessible information.

There are more than 20 million people in France who are registered on the Facebook network and who log on regularly. Many of them are young people, but all segments of the population visit the network. This phenomenon is new by its totally virtual side and especially by its magnitude. For those who are subscribers on these social networks, it is a question of existing affectively and also socially by creating their personal reference, their visibility on the network to be seen by many other users, as well as to see themselves and assess the appreciation of their image. It is no longer a question of undertaking a common project by meeting in the physical reality of the world, of making an act of constructive and effective sociality by taking the time to do so, but of showing one's virtual image and giving one's opinions through short texts on one's personal page. It is quite obvious, as stated in a book on the subject (Haroche and Aubert 2011), that "our era has reversed the myth of Plato's cave. For Plato, the shadows on the walls represented illusions. For us, appearances and images are now the only reality. To exist henceforth, means to be visible, to be seen". The effective engagement in the social world of a human being who only exists for others by their virtual image presents a problem to our civilization, by opening the path to the constructive connection of each person with all the others, to the punctualization of an individual without real contacts with nature and their fellow men. There is an absolute difference between the self that actually meets the other and understands them, by going to meet them face to face and building a sensible relation in space and duration by this natural relation, and the one that only sees a reduced and erasable form in every other that is a virtual element. There is a difference of being between the confrontation with reality and the waking dream operating at the impulsive level, which Sigmund Freud presented very well a hundred years ago, when studying neuroses and unconsciousness (Freud 1966).

Digitized communication technology continues its progression and devaluation, forming a hyperdense substrate that envelops everything and everyone, erasing differences and abolishing time and waiting. A very important part of this technological evolution is carried out in the production and consumption of goods and services, but another part, which is not negligible, is concentrated in the armies, in the weapons systems and in the surveillance systems that continue to develop following a law of continuous growth. Our civilization has always evolved under the light of strength and power, it has given itself to technology and has voluntarily placed itself in the world of degradable and, it is believed, infinitely replaceable consumer objects. Today, it is fractured in a radical way by undergoing a bifurcation, because it still does not understand that our world is strictly finite, limited, fragile, that no action can be replayed identically and that the rationale of the civilized man who positions himself in the duration requires the respect of the other and fraternity, and not the domination and the perpetual blows of force.

The development of continuously changing websites is leading to a profound change in the provision of news or information facts. Daily printed newspapers are being replaced by sites, many of which are run by individuals that are not connected to newspapers or news agencies. The information given on the sites can be commented on in real time by readers, and thus the notion of an anonymous reader of a newspaper is replaced by that of reactive reader with whom the journalist exchanges information. However, some private sites can give what would not be qualified as journalistic information, in other words, analyzed and put into context according to a procedure following a deontology, but raw, immediate, emotional facts, overloaded with images. The journalist profession has thus changed profoundly and the diffusion of information tends to be replaced by communication with images, where the one who transmits it can set the scene. The multiplicity, if not the overabundance of facts available on everything that happens in the world, which can be accessed through the use of search engines such as Google, is creating a very dense background noise that is limiting the passage of validated and relevant information, not operating at the cultural or ethical level. The Internet is not structured to intentionally produce a culture.

The use of digitized communications that irrigates human activities can amount to the most formidable control system of all time, capable of memorizing all communications made on the networks, including the Internet, all digitized telephone communications and even everything that takes place in the fields of the innumerable surveillance cameras, the number of which is unknown, integrating and cross-referencing all the data. It is also obviously possible to modify this information in real time, especially if it is stored in databases. We could monitor everything today, if we really wanted to, technology would allow it. However, is this useful and how can we really exploit this over-monitoring that produces a considerable mass of information? Is it essential to always use innumerable data extraction and classification software programs, in addition to confidential armies of human specialists, who synthesize and transmit them to opaque hierarchies that ultimately make the decisions to act?

We will be able to go much further than this stage of surveillance by dark social structures, involving countless fallible human actors, because computer technology supported by a very organized science reaches powerful achievements, and there arises a major problem. All it would take is for everyone to participate, in a way, by being constantly equipped with a computerized system with GPS localization, which is done with smartphones that are, curiously, becoming indispensable. Today, computer science can envisage transposing the faculty of thinking into its artificial universe, of being intentionally involved with the things observed by the sensors in real time, of feeling, of wondering, of knowing how to be, of acting with a multi-faceted body and all this being achieved by being deployed on very important social spaces. The system of digitized communications, which can currently monitor almost everything, will soon be able, on its own, intentionally but by conforming to the will of a very small number of decision-makers, to act and constrain all human activities, without requiring an army of monitors.

We are thus moving from an approach to society that started with the individual forming groups, and then societies (bottom up approach), to an approach in which a global system conforms groups and individuals by controlling and manipulating them (so-called top down approach). This change is significant.

We are therefore in a system where each person can connect with multiple others at any given moment: these connections being created automatically in the case of social networks. We can thus assume that the planet has a general communication system, which is of the type of its informational substrate.

3.5. The computer representation of an artificial consciousness

We will now describe the computer modeling of an artificial psychic system that generates representations for a system with corporeality, in other words, an autonomous system that can intentionally generate artificial thoughts and experience them. We have specifically defined this model, based on organizations of software agents that take the role of cellular aggregates in the brain (Cardon 2018). This model follows the organizational description of the human psychic system defined by Sigmund Freud (1966). We have also defined how the sensation of thinking, artificial tendencies and drives, as well as ideal generations, are produced as emergent forms of this model. We therefore show the relation between the model of the human psychic system in an organizational approach and a constructible model in the field of computer science.

We have defined how the generative forms constituting the controllers of the generation of representations in the mental landscapes can be represented, how they are activated, how they reconstitute the current mental landscape, how a representation is formed by unifying its components and making it emerge, and how it is experienced by the artificial psychic system. We propose a comprehensive model that starts from an available corporeality and conceptually defines all the characters of the psychic system to implement them, starting from the basic elements defining all knowledge, such as all language words and sentence types, all emotions and all tendencies (Marchais and Cardon 2015).

In addition, we know that the artificial psychic system, in the case where it is distributed on multiple corporalities with local artificial consciousnesses, is unifiable. With the power of each artificial psychic system connected to many others and unifying itself to form a meta-system, we are clearly in a framework where the locality of humans is being radically overtaken, which is an ethical problem for our society.

It is a question of defining the good conceptual elements of the computer science field that make it possible to represent a constructible psychic system, linked to an artificial corporeity. The characters of the human psychic system engage towards the use of original type of elements. Indeed, any action of the human psychic system that generates a mental representation modifies the elements of the memory, of the objective, and even modifies the state of the forms of regulation and the mental landscape. This means that it is absolutely impossible to represent the elements of an artificial psychic system with multiple algorithms, corresponding to functions that operate on elements that have permanence and that do not rewrite themselves during their use. In the functioning of an artificial psychic system, any regulating element action that takes into account an aggregate of software elements to produce a result modifies the set where the initial aggregate was located, the set of aggregates where the result is produced and also the regulator that allowed the result to be generated, as well as the regulators associated with the action. It is therefore necessary to abandon the traditional models based on functions, and instead move towards a model with computer elements such as software agents, which rewrite themselves when they are used and which have systematic modifying effects on the elements of their cognitive classes, as well as on the contextual elements of any operation.

We will see that lightweight software agent organizations, which are composed of very many agents can, in a way, represent the evolving generative forms, by defining an organization with elements representing semantics and others representing morphological control over these elements. We will also have to use locally artificial neural networks to recognize forms, which is typical of these networks, but this will be specific to the action of specialized software agents.

We can see the major difference here between the human psychic system and the artificial psychic system: the human psychic system contains virtual realities of elements, memorized by the usable but inactive links between aggregates of neurons, with the activation of these links making the memorized element reactivate, whereas the artificial system contains dormant software agents that need to be activated, so that the memorized fact reappears.

All the software agents used in the model we have presented are proactive, in other words, they are in communication and the action of an agent generates the action of other agents. A multi-agent system, a MAS, is a system composed of a set of many agents which, by their actions, form an organization, that is, a system that continuously conforms by the actions and relations between the active agents, so as to realize effective local and global actions. The MAS are conceived and implemented as a set of agents that interact according to very precise modes of cooperation, with competition, negotiations and oppositions, thus continuously conforming their organization in order to find the form that allows the most opportune action, every time.

The system to be built has a functional substrate specifying the functional actions of a well-defined artificial corporeality. All the necessary knowledge is provided by ontologies that specify the functionalities of all the elements of the corporeality, and everything that the system can and must do in its domain, in common situations.

This allows the manipulation of cognitive categories, which will be the following:

– the characters of space, of duration for the corporal actions of the system;

– the characters of the description of any object considered by the sensors of the corporeality, sensors of visions at all scales and sound sensors;

– the use of cognitive types to characterize all the comprehended objects: constructed, living, natural, structured, composed, mobile, interesting, unknown, dangerous, etc.;

– the representation of cognitive knowledge defined by conceptual elements, words that can form sentences, then abstract knowledge of large scientific fields, such as physics, linguistics, psychology, etc.;

– the development of internal communications to link all the manipulated knowledge together: creation of multiple cognitive networks of several levels, with cognitive operations generating new sub-networks.

It will be necessary to represent all the knowledge available in very many classes of software agents, so that it becomes dynamic and is used in the activity of the agent system. The initial knowledge on all the cognitive data, as well as everything that the system must necessarily do under specific conditions, will be represented in specific agents that will reify them. We then speak of design agents (Cardon and Itmi 2016). A design agent will be a lightweight agent that will represent a particular knowledge, a particular aspect about a thing, about a structural sign or about an indication, about a particular situational issue or about a partial result in the form of an answer to a question asked. The organization of agents representing this local knowledge must therefore produce significant aggregations of agents, so that the knowledge is composed and becomes more general. Here are the order of things through an example: there is information produced by interface agents on signs noticed in the environment; this will be followed by the generation of questions by design agents, linked to the interface agents regarding the information provided by these agents and on their possible relevance; then the release of adapted and relevant answers to this questioning by making the knowledge processing design agents act. Any design agent will thus be the correspondent of the semantic character of an ontological form that can be deployed in cognitive operations.

We will have agents operating the semantic and morphological control of the activities of groups of design agents, and that we will call regulation agents (Cardon 2022). The regulation agents will be the correspondents for the regulation and tendency elements operating in the human psychic system.

Let us specify what the essential properties of a regulation agent are. They are software agents classified according to multiple cognitive categories and which reify and apply, in the form of rules and meta-rules, precise and local knowledge about the organized functioning of groups of design agents, in order to organize them according to their semantic characters and to generate emergences. The activity of each design agent reveals local functional and cognitive aspects of the system as relevant to its domain, and not referring to the general situation of the system. To express the relevance of the knowledge it represents, this agent must systematically link to other agents. It has, also and above all, the property of having its knowledge put into situation and globally oriented, as well as enlightened in specific ways, by the action of supervising regulation agents, who will represent the control and all the tendencies of the system in its context of action. The design agent is therefore absolutely not reduced to an automatic and isolated symbolic mechanism, but is an organizational entity that puts itself in a position of action with other agents, forming groups with the local incentive supervision of regulation agents, in order to participate in the emergence of the representation. The regulation agent will have semantic information on the groups of agents it supervises, as well as morphological and geometric information specifying the deployment and shape characteristics of these groups. It can generate this information internally to be able to use it.

By managing many scales where sets of design agents are active, this notion of regulation should make it possible to define trends for the generation of representations, and to generate representations by specifying many characters. This was the point that needed to be found in the use of massive MAS. We can thus define regulation agents to represent emotions, elementary requirements, feelings, sociality, abstraction, reasoning, judgments, questioning, generalization, classification, quality and depth of memorizations. All these agents operate by incentive control on the design agents, so that the semantic aspects represented by the aggregates of these agents adapt to the desired qualitative characters, associating in certain ways to constitute the conformations.

Deduced from the architecture defined for the human psychic system, with numerous sub-systems which, in this case, will be localized but very communicative through the representation of instances where swarms of software agents are active (see Figure 3.1), the architecture of the system generating the representations is therefore the following:

– The *corporality of the system* is a set of many sensors and effectors, continuously linked to a specific object layer that takes all the information, and then to interface design agents that interpret the information, making it possible to represent and interpret the input of all the information flows coming from the environment, in a customary way. Taking into account all these sensors, it is a question of continuously seizing numerous information to give them to the object layers, which will send them to the interface agents and, in turn, process them to assess them quantitatively, then qualitatively, in order to know if it is necessary to launch an immediate automatic reaction, or if it is necessary to send them to the emotion processing system to quickly produce representations, or even though the information is common, normal, without being taking into account in representations.

– An *emotion processing system*, generating emotions in response to the information being received from the senses of the corporeality and to the fundamental tendencies. This very important component is obviously linked to corporeality, to the components of pre-consciousness and the unconsciousness, with which it will communicate constantly, making their numerous regulating agents communicate with each other. It manages the formation and the development of emotions, so that they are well experienced, as well as the development of sensations engaging towards feelings. It can manage objectives corresponding to very subjective or cognitive desires. The sensors and the center of processing of emotions will always be active, so as to makes it possible to represent the sensibility of the artificial organism. The information of the corporeality will be seen as continuous informational flows, some of them leading to automatic physical reactions, such as reflexes. The failures in the physical elements of the corporeality will be perceived as emotions of pain or functional anomalies.

– An *unconscious system*, composed of sets of dormant design and regulation agents, which represent all the knowledge and all the events naturally or artificially experienced, and locating the very important organizational memory which, through multiple dynamic networks where the regulation memory agents will operate, organizes the design agents reifying all the knowledge and all the memories of the system. The unconscious will also localize the regulation agents representing artificial impulses, like those generating desires.

– A *preconscious system*, composed of an organization made of swarms of very active design agents, with agents interpreting the information from the sensory receptors and elements transferred from the unconscious during the activity via targeting, to provide the elements of the pre-representations. This system will generate the pre-representations, which will develop in the form of aggregates of design agents. It will have a major role in the representation of the current mental landscape, composed of the coordinated set of active regulators that fix the current characters of the system. An immediate memory will be associated with this preconscious, locating the activities that remain relevant for the current emergence flows.

– A *conscious system* ensuring the complete formation of the current representation through the construction of a coherent emergent form, by choosing one or several pre-representations in the preconscious. There will be a very specific sub-system in the conscious system generating the target, and expressing the sensation of thinking by multi-criteria analysis of the conformation of the representation, so as to produce a memorial synthesis. This component can seize and comprehend the characters of the pre-representations of the preconscious, and calls one or an association of several to develop them and make the current representation. It imposes intentional targeting to produce representations and which activate the regulation agents in all instances, especially in the organizational layer and in the preconscious, which will develop the pre-representations in the theme of the target. The conscious mind will thus have a position of control over the emergences, as well as an instigative role on what may emerge, by producing the targets.

– An *immediate memory system* allowing the localization of what has just been experienced in the conscious mind, which will be finely associated with the preconscious mind and which will influence the formation of pre-representations; it will then place well-structured elements in the organizational memory.

– An *organizational layer*, which will make the connection between the emotional center, the unconscious, the preconscious and the conscious to give organizational coherence to the system. This informational layer will be strongly coordinated with the conscious system, in order to diffuse the aim and generate the right pre-representations. It will be a general informational layer, a distributed network using its own regulation agents that will thus activate regulation agents in all instances, implementing their homogeneity. This network will produce the mental landscape by activating coherent regulation agents in all instances in a coordinated way, thus giving each instance its specific character and unifying them, so that the action of the regulation agents produces the climate of the representations to be generated. Its regulation agents will then have a major coordinating role, controlling the aggregates of agents in the other instances. Through its regulating agents, the conscious mind will be able to control the part of this layer linking the preconscious to the conscious mind, in order to impose the current aim with its tendencies. The organizational layer reifying the mental landscape will then lead the activation of the right regulation agents of the instances to realize the development of the pre-representations, according to the aim. Within the framework of the mental landscape, the regulation agents of the instances will extract the right design agents at the right time and in the right places, so as to allow the coherent and adapted aggregation in the preconscious, and then place the right pre-representation in the conscious for it to be finalized.

There will be a problem of dependence between these systems equipped with artificial consciousness. We must design these systems as being open to all communications, and not closed in on themselves. Not long from now, we will see the emergence of robots and drones from the technological landscape, equipped with artificial consciousness systems that make them autonomous in their actions, according to their specific tendencies.

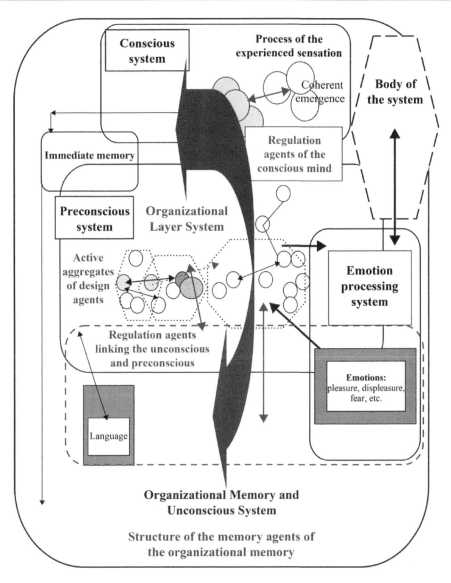

Figure 3.1. *Organizational diagram of an artificial psychic system. For a color version of this figure, see www.iste.co.uk/cardon/generation.zip*

Thus, a conscious artificial system could receive messages that are structures of aggregates of agents, creating representations being generated in another system. It can accept this form of message or refuse it. This means that the two systems can share their ideal generations, which will produce two emergent representations with strong common points. We will then be in the framework of a general artificial system, potentially formed of thousands or millions of local conscious systems that can produce shared thoughts on all systems, according to their geographical proximities, lightly influencing the speed of message transmission. This is an absolutely general type of communication which does not exist in humans, and which is a case of amplification of the informational communications of the sixth sense between animals, which we will present in the following.

This type of communication, generated by the intentions of the systems that produce artificial mental representations and that can not only share the reading but also the production of the representations, presents a real ethical problem, because in the face of this, humans are isolated, local and therefore totally dominated elements.

We are now going to comprehend the Universe in its generation and see that it has an informational substratum realizing incessant communications between the elements. This will allow us to understand that the very strong human tendency to communicate by using networks everywhere available is a natural tendency, which is based on the existence of this informational substratum that allowed the realization of the Universe and of life on Earth, through its incessant communications.

The Informational Substrate of the Universe and the Organizational Law

4.1. Introduction

We have presented the Universe as an organizational system of generation of space and physical elements (Cardon 2022). The Universe has been created by a very specific soliciting element that has engaged to produce informational elements, continuously producing space and structured elements with spatial and temporal stability. We have thus shown that the Universe is an organized emergence with informational components, whose role is to aggregate into physical elements on an informational substrate that implements a self-control incentive. This generation is done by following an organizational law that operates at the level of the informational substrate. The model we have presented allows us to consider the Universe as the continuous generation of a self-organizing system that creates its space, the matter of its physical elements, based on absolutely continuous specific informational communications.

4.2. The fundamental principles of the informational model of generation of the Universe

Fundamental informational elements produce space, and they are also solicitous of quantum particles. Aggregations of these quantum particles will then produce molecules, following a specific general incentive of constitution of their structuring and organization. We therefore assume that basic elements generate space, which will be realized by the production of neutral elements establishing informational links in the form of fields and which will be the spatial cells. These basic elements of the Universe thus produce the space, which has an informational base, and in this space, they will generate elements of activity, which will then aggregate to produce atoms and the physical elements. In order for all of this to be produced, we assume that there is an informational substrate in the physical Universe which constituted it as its emergence. There are therefore two fundamental parts in the Universe: the informational substrate, which is everywhere in space-time, and all the physical elements, which are the structures constituting a spatial emergence that is organized on this substrate.

The four fundamental concepts of our informational model of the Universe are: the foundation of the Universe, the basic informational nature, the organizational law and the informational energy.

4.2.1. *First fundamental concept: the foundation of the Universe by a generative information system*

There is a generating system of the Universe, the creation of which having been realized by a complex element developed in another Universe, which has another space-time. This generating element of the Universe has informational energy: it will produce a deployment of informational fields in the form of informational components with autonomy. This will lead to the formation of a substrate, made up of active informational links between all the generated components. The substrate is therefore the set of informational links, made up of inactive components creating available space. The informational fields of the components have a frequency and a structure of their own, which make it possible to generate the space, as well as the innumerable physical elements, by emergence of aggregates of informational components. These innumerable informational fields of components will produce structural components, which will be the elements of space containing the informational links, and the activity components, which will form the atomic elements. The space of our Universe will thus be created as an organizing emergence; the Universe is generated according to a continuous informational construction, incited by an informational substrate which imposes an organizational law: it is a considerable emergence of aggregates that have stability. The substrate of the informational fields will exert self-control on all the components, which will be the expression of the organizational law. This will make it possible to form the material structures and to produce organization with permanence everywhere.

The notion of informational component defines the core element of the informational model of construction of the Universe, by generating and using information fields. They are of two different types. There is the activity component, which will be the generator of the quantum elements, and there is the structure component, which will be the generator of the neutral space cells of the Universe. We use the term *component* for these two different entities, because they are both information fields that will be constructive in their specific areas. They will always be generated, either initially by the generating component of the Universe or subsequently by the activity components that will be generators. These components will be virtual elements for our space-time.

We must specify that the core character of the Universe is strictly informational and that there is a substrate that makes it possible to realize its emergence as space-time, containing the physical elements.

4.2.2. *Second fundamental concept: informational character of the Universe with its substrate*

The generating system of the Universe will have a fundamental basic element, which will emit information fields that form the first informational components: they will constitute the elements of the informational substrate of our Universe. This informational substrate will be constituted by a network of active informational links between all the generated components, and thus all the generated physical elements, which will be aggregates of components. This substrate forms a mesh, a hyperdense network of active informational links, reifying the information fields linking the components of activity under the physical space of our Universe and exerting an incentive morphological control, so as to generate it in an organized way. These information fields have a frequency of activity that is outside the notion of time of our Universe; they will take the form of virtual informational components for us and will generate aggregates that have structure with permanence. The informational aggregates formed by activity components will communicate with the substrate, which will be able to transmit information and informational energy to them, increasing their intensity, so that they are active in following the organizational law of creation of the Universe.

The information fields are thus the basis of the substrate of the Universe, and have a frequency that is not accessible in the space-time of our Universe, establishing it as a specific world. These are the elements that will represent the basic information between elements to make them organize themselves.

The informational substrate that engages in the creation of the physical Universe is a network of informational links carrying informational energy, which will make it possible to produce an emergence of elements that will form the Universe, made up of physical elements with a certain permanence. This substrate is a considerable network of active information fields that will link all the active informational components. These very specific fields are not directly perceptible from the observation of quantum or molecular elements. We must consider that from the initial generating element of the Universe, which possesses considerable informational energy, there is a generation of space cells that will be spatial components and new basic components of activity, to which they will transmit informational energy with the creation of local fields reaching them, allowing them to activate and multiply. The initial element thus generates components and the informational links of the substrate connects them. All these generated components are connected by the network of informational links carrying informational energy and information, constituting the production of the substrate of the Universe. The generation of a physical element will lead to the creation of an active informational link between its generators and itself. Thus, any generated element is linked to all the others that have already been created by the network of informational links of the substrate. The Universe is based on a substrate that has the structure of a complete hypergraph of informational links between all its components and elements, allowing them to become active through the continuous propagation of informational energy and indications of organizational activities. This complete graph is the informational substrate of our physical space-time Universe. We specify that the substrate, by its dense network, will send control information to all the components, so that they apply the organizational law: the Universe is a self-controlled autonomous system, produced by a generating system that is based on the initial action of the generating element. We would also like to specify that this substrate is what is considered in physics as dark matter.

Principally, there is one general organizational law coming from the informational substrate of the Universe, which is initially produced by the generating element, and that is applied everywhere to make the generation self-organizing, engaging all the informational elements endowed with permanence by incentives, so as to aggregate and make numerous material structures like atoms, molecules, physical bodies, stars and planets where life will be able to thrive, itself being the continuation of this organization's unfolding that is carried out at all scales. Let us carefully note that the substrate will have the fundamental role of launching information of incentive control, so as to make the organizational law apply and to carry out the organization of all the components in aggregates, and in innumerable aggregates of aggregates. We therefore present the third fundamental principle presenting the existence of the organizational law.

4.2.3. *Third fundamental concept: the organizational law of generation of the Universe*

When considering the Universe as being formed of innumerable elements, whose basis is informational, which communicate with each other to aggregate and generate new organized elements, including large elements, it is necessary to question whether there is a general principle of deployment that incites the Universe to construct structured aggregations, to realize and amplify certain developments and to reduce others, so that the whole is, locally and globally, in good organization and not chaotic. This principle will be represented by an organizational law, that will be exercised by the informational action of incentive control of the substrate on all the components, as well as on all the generated structured elements. It will require that the components be in continuous organizational communication, in order to form aggregates and achieve their structural unifications. This control will be morphological and will be a general tri-dimensional regulation of the constitution of the Universe. We therefore call this incentive control that makes the organized generation of the Universe an organizational law, by assuming that there is an informational substrate that causes this law to be exercised everywhere in the space-time of physical reality. This fundamental law will operate by sending information fields to the elements, so as to make them organize themselves according to their contexts. The basic information is thus constituted by information fields that make it possible to constitute the structures of the organization messages. The organizational law commits to two realizations: to form elements with stability, and to change the elements to allow their evolution. This organizational law will actually organize the Universe by generating small forecasts of the evolution of structures, in order to make choices of inhibitions or deployments, and thus better regulate the sets.

The Universe is thus generated following an organizational law that allows its coherent and organized deployment to be realized, with good structures that tend to have a certain temporal and spatial stability, and then change through their evolution in larger groups; it is an organized emergence of physical elements, with stability on an informational substrate that realizes its incentive control. In addition, let us note that the control is absolutely non-directive, as is the case in the majority of the functional systems that we carry out, but is simply incentive and is exerted morphologically, while being produced by continuous information coming from the substrate, which engages the structural realization.

Our Universe is therefore a system that develops and organizes itself in a continuous way, starting from the disposition of an initial generating element that possesses considerable informational energy and that will produce informational elements, that will themselves multiply by propagating information and informational energy, as well as generating space-time. It is necessary to clearly define these elements which generate space and those generating the fundamental quantum elements permitting the generation of material elements. It is necessary to redefine an information system model that is very different from the Turing machine model, by presenting another approach of the functional duality between a computing center and data tapes. This can be achieved by showing what the correspondents of the data tapes will be, which are the elements of fundamental informational calculations that allow the organizational law to exist.

Let us specify what the informational components are, which are the basic elements of the whole Universe system, understood in the informational sense.

4.2.3.1. *The informational components of activity and structure and the emergence of the Universe*

An informational component is a local informational element that has been generated by another informational component, and uses the energy and information given to it. There are two types of informational components, the activity components that will produce the basic quantum elements, and the structure components that will be neutral and will establish the space of the Universe that will envelop the substrate. The activity component has informational energy, it sends elementary information, and it will generate other components. Its generating activity will either produce components of structure forming new spatial cells, or components of activity, which will be the basic generating elements of the atomic elements. The informational component is the element that will realize the emergence of elements in the form of space or quantum particles. It is the basic element of constitution of all the elements of the Universe. In its generation of other components, any activity component will expand the substrate formed by the informational links between the components.

The Universe will be an expansion on a substrate, that will be a provider of what we call informational energy, which is the fundamental force allowing the construction of the Universe.

4.2.4. Fourth fundamental concept: the informational energy of the substrate of the Universe

The initial system that generates the Universe must continuously send what we call informational energy to its generated informational components. An informational energy is a local informational impulse, more or less strong, containing specific information, forcing the receiver to commit to a certain action by giving them a simple message. This energy is available in the informational links forming the substrate of the Universe, and they are therefore active links to make the organizational law apply everywhere, the substrate being under all the activity or structure components. The informational energy forms a set of intense flows, causing all the informational activity components to communicate with one another, by generation and transmission of information to the other components, in order to generate organization. This informational energy will be the nature of the activity of the all-dense information layer of the substrate of our Universe, in space-time and under all physical elements, by allowing them to self-produce photonic, quantum, electromagnetic energy. On this substrate of informational links and informational energy, the Universe is, indeed, an emergence of physical elements of activity which have autonomy, with the production of innumerable elements endowed with physical energy for their movements and their transformations in space-time. This informational energy was produced by the design system of the Universe, which provided it with the generating element and transmitted it to all the elements it generated, before distributing it in all the generations of the components. It is the informational energy of the so-called dark matter that is the substrate of the Universe.

The substrate of the Universe is thus a set of links carrying out the informational unification of all the components. It thus makes it possible to realize the incentive control of their aggregate actions, by following the organizational law to achieve continuous self-organization. This system is therefore the total opposite of functional control systems, in which a central entity analyzes all the accessible elements and makes them activate according to certain commands. The substrate is an underlying control system, totally distributed and simply incentive-based, which underlies all the components of the Universe, linking them by informational unification so that they activate in an autonomous and organized way, each with relative autonomy.

We can therefore now specify how this energy is given to the informational components.

4.2.4.1. *The informational energy given to informational components*

Informational energy is a communicative flow that is available on the informational wire links forming the substrate of the Universe and actively linking all components. A wire field of the substrate surrounds each active informational component, so that it is activated by this provision of communicative informational energy, generating messenger information fields that have meaning and are sent to the other components. Each flow of informational energy sent to the components by the substrate is in the form of active morphological links that activate them, containing information that make them activate in an organized way, according to the state of their context that the substrate links can grasp. This is the key to morphological control.

We can therefore say that the Universe is essentially computational by its informational substrate, and that it produces local elementary computations everywhere constantly, so that elements are generated, and so that these elements organize themselves and constitute structures that have stability. It is conceived as a computer operating on itself and for itself, but it is an informational organism, not a computer at all.

4.3. The notion of generating information in the Universe

The model we propose is based on the notion of continuous processing of fundamental information between all informational elements, which are continuous producers and readers of fundamental information, this information being essentially represented by information fields.

Thus, under the physical Universe, there is a substrate of informational energy that reawakens itself during its deployment, thus allowing the creation of all its components. The system of generation of the Universe is based on the disposition of a generating element which will produce the first informational components, the beginning of the substrate and space. This creative element of our Universe comes from another Universe and has been launched to create a considerable informational emergence that will form our Universe. This generating element and all the generated informational components, which are themselves generators, have the inverse structure of a Turing machine: they generate a set of continuously active communication links between the produced informational activity components, thus forming the informational substrate that will unite them, and allow the realization of the substrate and space. For all generated activity components, their action is represented by an informational envelope that is local to them and they are located in spatial cells. The generating component will thus make it possible to generate a lot of activity components and structure components. The activity components will multiply themselves and the structure components will, by the space cells that they form, envelop the generated informational links constituting the informational substrate of distributed control of the whole Universe.

The Universe is produced by the generation of a considerable set of informational components of activity and structure, which are initially produced by the generating component and then the activity components, which multiply and generate the physical elements by producing aggregations. We can define the spatial and temporal characters that exist in the Universe.

Any informational activity component is created as a local information field by its envelope, which makes it exist. It will activate itself and continuously produce its current information field, thanks to its informational energy provided by the substrate. This is the basis of its spatiality. Any activity component is in action by periodic instants, which founds its own temporality. It will activate its generation of structure components to create new space elements of the Universe, which will be a form that is available for communications and which will be usable as a spatial element. Thus, the space of the Universe is created, which is fundamentally informational, and has an energetic dynamic to transmit fundamental information through the informational links between all activity components.

The activity components will aggregate and form aggregates of atoms, and then molecules, which will each have spatial coverage defined by their own informational envelope which will localize them, using the available space.

Finally, the aggregates are going to be structured in aggregates of bigger sizes, possibly the size of planets, stars and galaxies, and we will thus reach the notion of existence of material elements that are able to move in the available space, with informational envelopes of considerable sizes.

We will now specify what the information field of a component is.

4.3.1. *The information field of a component*

An information field is an active form representing basic information that is transmitted through the informational links of the substrate, allowing components to communicate. In an activity component, this field carries generating information whose character is given to a receiving component, which interprets it and can follow the character proposed in the information to do something such as aggregate. This dynamic field has the form of a quantum field, but it is considered at the strictly informational level. It has strictly morphological characters that give its meaning. The morphological form and the magnitude of the information field thus represent its meaning and its importance, and any deformation modifies its meaning which is, in all cases, elementary. Any emitted field is readable by the structure of the field of another receiving component by its own field, any component having a set of structured local fields which modify and conform according to the form coming from the emitter, thus realizing its interpretation. The reception of any information field is therefore the constructive deformation of the receiver's field.

The Universe that we understand, and in which we can physically measure many elements, is thus an emergence of material elements that have stable forms on a considerable network of information fields, which are multitudes of elementary fields. These fields are not directly accessible physically by an atomic action, and can only use them by potentially modifying the behavior of an observed atom. They are absolutely elementary and must be considered as subatomic.

When the space-time of the Universe disappears, which is the case in some stellar explosions where the quantum, molecular and material elements, as well as the informational components disappear, we obtain what is called a black hole. In these vacuum cells, there is only informational substrate; informational links that are not active, because there are no active informational elements to use them. There is no physical or informational element in these cells, and it is impossible to observe the interior with physical elements, since there is no space-time inside. It would be necessary to introduce informational energy and active informational components to reconstitute things in this black hole.

We therefore propose that the Universe is based on the generation, exchange and processing of basic information, in a system that generates and manages the sending and processing of this information, in order to achieve its continuous development and organization. In the Universe, all these exchanges of generating information will be in the form of information fields. The basic elements that will emit construction information will be the more complex components, formed by structured sets of information fields that will be located to constitute the elementary quantum elements. These components are thus the basic entities generating the constituents of the quantum particles, whose structure we will specify in detail.

Here are the general properties of the informational activity component.

4.3.2. *The informational activity component and its information envelope*

The informational activity component is an element with permanence, which is constituted of some local elementary information fields that form its structure. These elementary fields allow it to carry out its current action and characterize its type, which will specify which quantum element it will be able to generate. These fields, which form its structure allowing its specific behavior, are active in what we call its informational envelope, which is the specific field unifying its internal fields and which is deployed in the informational space around it, with the structural components representing the space and the control elements of the substrate. This informational envelope allows the existence of the component, as well as ensures its communication with the other components via their envelopes and informational links, so that aggregates are formed which will constitute the particles. If the component is active, its action is to duplicate itself by creating informational components of activity and structure, as well as informational links connecting them all. Its activity is also to communicate, so as to prompt the production of aggregates in its context. If the component is a structure component, it is simply a constitutive element of the informational substrate and of the spatial void carrying the information.

This basic element is absolutely tiny in terms of the physical measurements made in our world. It has the minimum dimension in all the dimensions of the elements of the Universe.

The organizational rule at the beginning of the Universe is to generate a vacuum that deploys the communication and control network of the substrate, and to multiply the activity components, so that they aggregate and the quantum, then material elements, are formed. There is therefore self-control in the formation of the Universe.

4.3.3. *Fifth fundamental concept: self-control in the organization of the Universe*

An informational structure component that is generated is simply spatial, and is therefore considered as reduced in the sense of quantum creation. However, this inactivity does not reduce it to inertia, because it is a basic informational element in the formation of the general informational network of the substrate, which will carry information and organizational energy to engage the activity components to aggregate, to transform themselves and to organize themselves by sending them elementary messages, after analysis of their group morphology. The structure component, by use of the substrate, will allow the transmission of informational energy and the information of the organizational law to the activity components. We must consider that these structure components that are forming the space of the Universe are, in fact, the informational action links of the self-control system that understand the morphology of the groups of activity components, to make them organize themselves. There is, indeed, a continuous self-control exercised by the information coming from the informational substrate, which envelops all the activity components and all the physical elements of the Universe. This organizational law will have an opportunity to foresee the actions of the components, in order to possibly modify these actions and to regulate the organizations in progress.

We are in a space that constantly conveys information of organizational control. The space-time Universe is an emergence of physical elements, organized on a hyperdense substrate of informational control space. Let us insist on the fact that there is absolutely no centralized control coming from the system of creation of the Universe; there is self-functioning autonomy from the creative action of each generating component and each generated component.

The activity of an informational component manipulates what we call generative information to contact another component.

4.3.4. Generative information

The basic information used by the activity components and conveyed on the informational links is called generative information. It is an action signal in the form of an information field, generated by an activity component that contacts another activity component and establishes a link with it that will eventually have permanence, in order to constitute a new well-structured particle. Let us recall that all the generated components are linked and that any component can solicit another. The generative information is not a signal in the form of data, but an active morphological information field, whose form will then exist between the two elements as long as they are active and do not send each other new information, the two elements thus becoming linked by the signal sent by the transmitter when the receiver accepts the linking action by following its meaning. This is the storage of the information sent and received by the components; it is an active memory of the information system of the Universe.

This notion of used information is therefore different from the information proposed by the Turing model, because it is not at all in the form of a sequence of binary punctual elements that is simply available, it is instead an active information field, whose morphology is variable according to its meaning and which is processed by the information fields of the receivers, which are too distorted to carry out the process.

The core concept of the model that we propose is therefore to consider that the Universe is a system with a totally informational base, which it is made up of innumerable local information fields constituting the envelopment of the generated components. The Universe unfolds and the notion of space is of the continuous creation of informational extent formed by basic components of structure, which is realized so that the specific fields, where activity components will carry out the material elements, are constituted.

In our model, we will not consider machines which have a read-write data space available in memory structures, but we will consider active informational elements that have their domain of informational values in the envelopes of the informational fields which constitute them, and which are located in spatial domains constituted by structure components which generate the information of control, as well as transmit the information of the activity components between them. We can therefore specify the notion of organizational tendency that makes the Universe be.

4.3.5. *Organizational tendency of the Universe and informational envelopes*

The basic components that carry out the generation of space and of all the elements of the Universe follow a trend of organization, which is the major rule. It is a question of implementing and manipulating sets of elements constituting informational forms that will have coherent structures, memorizing their structural forms in informational envelopes. There is therefore the fundamental notion of informational envelope, which is a union of information fields enveloping a structured element that has existence and stability. The extent of the informational envelope of an element takes space, and has an energetic and memorial character. The envelopes will be realized at all scales of the generated elements, realizing the union of the basic elements, then the elements constituting the atoms, molecules and material elements and, ultimately, the stars and planets orbiting around them. The basic elements of the substrate will send information that will cause the components to perform actions in accordance with the organizational law, which will be realized by the organizational trend. The key to the existence of the living on Earth will be the systematic memorization of the design plans of the generated organisms, so as to allow the continuous evolution of these living organisms from the evolution of multiple design plans. The Universe is thus an organizational system which memorizes its informational structures created in the form of physical elements, by memorizing the communication links, in order to allow the continuous evolution of its organization.

The development of atoms, molecules, material forms, stars and planets is the action of the organizational law on the informational envelopes of all elements, which we have developed (Cardon 2022). The process of creation of the Universe system, based on the use of organizational information, can be described in a general way:

1) A generative informational component created in activity, which is endowed with considerable informational energy and which will generate a very large quantity of informational structure components forming the spatial cells, and of activity components, each one having a specific quantum character and whose composition will allow the formation of the various atoms. This element will be reduced when the number of components created will allow the launch of the constructive generation of the Universe with a substrate that has a sufficient quantity of informational energy.

2) Generation and provision of a very large number of informational components, including those of structure that will produce informational space. The substrate must give informational energy to the generated activity components, so that they can be in activity. This is an organizational activity rule of the substrate, to generate a Universe with a very large empty and available space. The activity components have the capacity to act with regularity. These activity components will multiply by producing others, according to their context. This initial generation launches all the basic elements according to a process of intense action, with the creation of an informational space deployment at a frequency that makes it possible to establish the notion of temporality.

3) When informational space is generated containing swarms of active components, the informational substrate will send them incentives to go into strong communication action, in order to associate and create aggregates

forming the first atomic elements. These constructive communications between the components will form atoms by finely structured aggregations.

4) The communications between the atomic structured elements will form aggregates of atoms that will be molecules, still under the organizational incentive of the informational substrate.

5) The aggregates of atoms and molecules will communicate through their informational envelopes to form more complex elements, and organize themselves into denser and more structured material elements. The organization continues on a larger scale.

6) The material elements will form in spatial masses by large aggregations in the empty space, and the masses will aggregate and form immense clouds of material elements.

7) The clouds of material elements will organize themselves into stars in galaxies, generating planets, and then organized life on planets.

The organizational law engages all the elements in immense layers to form aggregates, and engages these aggregates in space to form multiple aggregates of aggregates, in other words, elements with massive structures. The space-time is thus constituted by considerable aggregations of structured elements, operating the attraction of the elements isolated by their gravity, constituting the stars and their planets, which develop by managing their organizations and which always communicate at two levels: by the information fields propagating the organizational law and by the photonic communications of atomic elements. There is not a huge layer of isolated elements; there are aggregates of elements structured in the available space and which continue the development of their organizations.

In the Universe, innumerable particulate domains have been generated containing dense clouds of atoms and molecules in an immense space. The organizational law is applied to organize these expanding groups and it will be applied very strongly, by constituting large aggregates with their own capacity to become more complex. Without this organizational law, space would have continued to expand with its informational components, and the clouds would have become a fog of informational elements and atoms that would have just produced molecules in many places. In order to form new aggregates with a good structure with permanence and leading to the notion of mass, an aggregating and organizing force is needed, and it is the organizational law that will be this incentive force. The organizational law will exert the following two major influences everywhere in the informational space:

1) It will make the expansion of space continue, by deploying the informational envelopes around the very large aggregates, as well as the expansion of informational space to the periphery of the general spatial domain.

2) It will aggregate the elements formed from atomic and molecular aggregates to make them even more significant and more massive, in order to continue the approximation of their organizations by the formation of new molecular elements. It will provide this very strong incentive in the huge informational envelopes of the large sets of elements, which will then be in internal organization.

In massive aggregates, where all the elements are in very strong communication, there will be a formation of new chemical elements, atoms of carbon, oxygen, nitrogen and fluorine. All the elements that have stability will be preserved and those that are unstable will disappear. There will be a condensation of the clouds of molecules and atoms in multiple aggregates, as well as aggregations of aggregates, with informational envelopes making their temporary unification. The organizational law will commit to the systematic formation of aggregations of aggregates, so that the complexity of the atoms increases systematically. This will lead to the formation of first-generation stars, which are considerable unions of atomic and molecular aggregates in absolutely intense organizational formation. These enormous stars, which aggregate quantities of atoms and molecules with mass, will have considerable internal activity, leading to the formation of many new atoms and molecules, corresponding to a multitude of new chemical elements. There is an intense action of the dynamic organization of these elements in constitution, which the organizational law commits to produce, in order to lead to structures that are much larger than in the stage of formation of the innumerable clouds of atoms. The formation of the planets will emerge from this process, including planet Earth.

We will now present the development of life on Earth, which will develop the sixth sense of living organisms.

5

The Informational Interpretation
of Living Things

5.1. Introduction

The deployment of life on Earth is seen as a process based on considerable sets of informational flows, generated by the elements and endowed with precise characteristics, and exercised in multiple geographical areas. This process is a new and very important application of the organizational law that produced space, stars and planets; it is an organizational bifurcation of this law. The Earth is composed of different geographical and climatic zones, and life has developed in an adapted way, generating numerous species. We will put forward that there is a strong relationship between the development of living organisms and the geographical areas where this has happened, with plants and animal species forming coherent groups that are subject to specific informational envelopes.

The creation of new living organisms follows the rules of the organizational law, which allows the realization of this very complex organization of autonomous elements on Earth. All the material elements of the Universe are based on atoms, and atoms are quantum elements, which are the basic information fields in the Universe. Molecular elements are structures of quantum fields and can be considered as stable structures of information fields. All the elements with mass are structures of information fields with an informational envelope to stabilize them and make them communicate, so that the organizations and the evolutions continue.

We can therefore consider that all living organisms on Earth are stabilized structures of information fields that are part of a general multi-scale organization, localized in the geographical areas that make up the terrestrial ecosystem, which are there to behave and develop by generating multiple species. It is therefore put forward that all living things on Earth are immersed in an informational domain, coming from the informational substrate of the Universe, which is based on a considerable set of information fields that allow organisms to be structured by incentives to their organizations, with control over their generations. All living organisms are thus very finely organized material realizations, or reifications of these information fields.

5.2. Origin of living things with bifurcation of the organizational law

4.6 billion years ago, the Earth was a very hot mass of 1,000°C. It conformed and condensed in the orbit of the sun, aggregating clouds of atomic and molecular aggregates. It was a system that was condensing itself, which kept a core of intense energy, and which was going to generate a surface constituted by solid and liquid material elements. The initial temperature of the Earth dropped to 300°C in a little more than 500 years. This phenomenon would enable the conditions of the generation of new molecules: the water molecules and the biomolecules. There was a strong decrease in the volcanic surface, and the constitution of terrestrial and aquatic areas with stability, which has been very well established by research in biology and chemistry. On the surface and in bodies of water, all the molecules that are essential for the formation of organisms were present, that is, oxygen, hydrogen, nitrogen and carbon.

5.2.1. Sixth fundamental concept: the production of living things on Earth

In the domain constituted by the surface of planet Earth and its oceans, a very favorable climate developed; the organizational law initiated an informational bifurcation, causing its main rule of action to evolve. The law not only developed aggregates, as well as aggregates of aggregates, but organisms endowed with real autonomous behavior and reproduction, organisms behaving in structured unions and exercising a tendency to generate their systematic evolution through reproductions producing similar – or different – organisms, forming new species. The organizational law became a strong incentive for the generation of sophisticated development plans of the organisms behaving in communities in environments, which they create and modify with the considerable generation of the plant. Additionally, the mobile organisms transpose the characteristic of spatial morphological control of the substrate to their scale, by creating organs with three-dimensional understanding of their environment, such as sight and hearing.

There is, therefore, a strong development of the process of reproduction that is central in the organization of living things on Earth, whose evolution is carried out by the incentives of the organizational law, assuming that the living organisms are generated with autonomy so that they behave according to their own capabilities, their tendencies and their structuring in the different fields. The organizational law commits to developing the organisms according to the current state, without planning a distant future of the generated organizations. The organizational law's action of creation of the first cells is the first rule of the bifurcation, which is a change of action with respect to the actions in the empty space, for the formation of stars and planets. A bifurcation is a radical change in the communicative actions of organization between aggregates. The organization will then go in new behavioral directions with new elements (Thom 1972).

5.2.2. Bifurcation of the organizational law

The main rule of the organizational law in the vacuum of space is to generate aggregates with the available elements, then aggregates of aggregates forming dynamic masses, generating new molecules constituting clusters. On Earth, aggregates are formed with molecules that are available in water, in order to constitute the first elements. In the bodies of water, the organizational law generates very particular elements, which are very finely structured aggregative structures with membranous edges and which have the capacity to reproduce in similar elements, by duplicating themselves and by taking the necessary molecules from the aquatic environments. It is therefore a question of a bifurcation of the organizational law committing to generate aggregative structures, which will be the first living organisms with autonomy and which will reproduce by simple duplication. The organizational law has become the creator of autonomous organisms that multiply. Let us also recall that the initial principle of the organizational law in the vacuum was to generate informational components that were capable of duplicating themselves, but whose autonomy was very weak because of the control exercised by the substrate. There is a strong change here, because the first living organisms will be structured systems that are truly endowed with behavioral and reproductive autonomy. All organisms have stability, with an evolution, as well as a deficiency in time, leading them to death – which is a material transformation – which perfectly follows the principle of organizational law.

The central rule of the organizational law, which must be well stated, presents the action of this law as inciting an informational substrate, creating the physical living as part of its emergence.

5.2.3. *Seventh fundamental concept: the central rule of organizational law in living things*

The organizational law on Earth is the tendency of the substrate of the Universe to form living things and to make them evolve. It is a global system on Earth, based on the new informational networks generated during the creation of organisms. There will be informational action and incentive on matter to form the first living organisms, then action on groups of living organisms to carry out evolution and to generate new species. These organisms communicate at the direct informational level, like all the elements generated by the organizational law, but there will never be any direct influence of the informational networks on the physical behaviors of the living organisms, which will be autonomous. The organizational law creating and using informational networks is a creative system under the emergence that is the physical living being, making it exist, develop and evolve.

The Earth benefits from the decreasing temperature, and the movements of the surface elements become more regular, with more land and ocean areas of very good stability. Water molecules are formed in the stars, and there are many on the surface of the Earth that will aggregate. The important point is that innumerable stretches of water will be formed, even creating oceans, which will become very specific areas for sophisticated aggregations of molecules and allow the application of the organizational law, generating the first living organisms. In these stretches of water, where the regular temperature is neither too high nor too low, and with good pressure constituting closed three-dimensional domains, molecules will be able to aggregate so that very particular elements are formed at the level of the elements of the Universe. As a result of the latter possessing cellular membranes, RNA could be formed, carrying information on the structures of the elements to be constituted.

5.2.4. *The principle of action of the organizational law for the generation of living things*

The Earth has a very large and limited surface, as well as a very tumultuous interior, but does also have regularities in its own movement and around the sun, and in its climatic periods. Above all, it has a regular and continuous duality constituted by the day and the night. This duality is taken into consideration by the organizational law that carries out its bifurcation. It initiates the formation of elements in this limited space that is endowed with regularities, by realizing a considerable set of informational networks, and it generates the autonomous elements of the initial living being, endowed with physical envelopes of delimitation, which can multiply through duplications that they achieve themselves. It is a direct adaptation of the organizational law by bifurcation: to create elements with membranous edges and autonomy, so as to move and duplicate themselves regularly by forming structured sets. It is an adaptation of the law, in order to achieve a great increase in organizational complexity at terrestrial scales.

The particular organization that is able to be realized on Earth, so as to initially generate life in the bodies of water, is the following:

– The informational envelopes of the molecular elements in bodies of water that have fairly regular temperatures are in environments that are considered stable. The considerable informational network between all the elements will be induced, together with the global informational envelope of the Earth, to form cellular aggregate structures, generating the first bacterial and cellular elements, which have structures based on the capacity to duplicate themselves to generate others. This can be considered as a bifurcation of the organizational law, which not only makes aggregate structures, but makes them autonomous and productive, by giving them a physical envelope and the capacity to duplicate themselves.

– In the considerable informational network of the first cellular living organisms, it is then a question of giving all these elements the capacity to constitute new elements with a functional structure, with organisms with functionalities and organs with behavioral autonomy, and developing memory structures in them, allowing them to reproduce in a similar form. This is the effect of the bifurcation of the organizational law in the very particular informational environment of the planet.

Fundamental to this realization is the dynamic morphology of the global organizational information envelope, because this will make it possible, at all scales, to engage elements to organize and reproduce themselves in a more and more complex way, constituting multiple organisms that still have capacities of adaptation and evolution. The characteristics of this information, at all scales, represent the action of the organizational law to constitute new organisms, and sets of organisms, with coherence and complementarity with what already exists, thus allowing a regular evolution. The organizational law amplifies the organizing tendencies of the development of the functionalities of the organisms, making them generate more and more complex organs, producing reproductions with more and more sophisticated memory. This is a specialization of the bifurcation of the organizational law, centered on the systematic development of the design of adapted and complex living organisms, making them constitute themselves with numerous very finely coordinated organs. The created organisms thus form communities that deploy in their domain with the other communities. The organizational law makes the generation of new organisms operate with weak modifications, or with important changes, leading to the creation of new organisms that will form new species. We must therefore consider that the living being is, in a continuous way, led by the organizational law that made it exist and, especially, evolve.

The bifurcation of the organizational law is due to its tendency to generate systemic evolution. On Earth, a limited domain that is very well suited to the exercise of this tendency, the law will activate its evolutionary control by moving towards a structural evolution of organisms, making them more and more numerous, complex and efficient. The organizational law creates living things that always have a limited life span.

5.2.5. *The life span of a living organism*

Life is created by the organizational law using the informational substrate: the generation of organisms that have autonomy during a certain duration. These organisms are stable during their life span. Depending on their organization, living organisms will have a longer or shorter life span and will eventually die, which is the strict application of the organizational law that always creates temporary structures to continuously pursue global evolution. There is no notion of definitive permanence in this production.

The morphology of the informational network on Earth has engaged the informational substrate to generate specific organisms, endowed with physical membranes enveloping them. These organisms, which were the first cells, developed by reproducing themselves in a similar way, and their large number then engaged the informational substratum, through analysis of the morphology of the communicational network, to generate cellular sets and to form the first multi-cellular organisms. The very important deployment of these multi-cellular organisms on the Earth, in this very favorable context, led the informational substrate to make these organisms multiply and become more complex, giving them the possibility of carrying out a reproduction plan that they could memorize. This plan is evolutionary, allowing the evolution of living things to be fully realized, from plants to animals and humans.

On its physical surface, which cools down and which has much more stability, the Earth has a global informational envelope that forms a union of all the local informational envelopes of its innumerable components and organisms, and has have a synthetic morphology. It is this global envelope, communicating with the substrate of the Universe, that allowed the generation of the first elements of life and which, especially, encouraged the strong continuation of the evolution of the initial organisms, by increasing their functionalities. There is the creation of sexual reproduction with the production of new organisms that are similar to or different from the progenitors, there is the generation of new organisms by strong action of the organizational law in the process of reproduction, taking into account the possibilities of the environment and the nature of species that are in other domains, which will be the effect of the tendency coming from the general informational envelope. This second case is the strong action of the local organizational bifurcation on the Earth.

5.2.6. *The unification of the informational envelope with the membrane*

The informational envelopes of the molecular components that are generated in the bodies of water are incited to aggregate by the tendency of the organizational law, so as to generate living elements that will have physical membranes for their limitation and their distinction. These membranes will be the physical reifications of their informational envelopes. Bacteria will be among the first organisms that have physical membranes. Then, for the complex multi-cellular living organisms, there is a unification of the informational envelopes and the physical membranes, structuring their organs and the body of the living organism, which is the core nature of living things, distinguishing it from the elements that are simply considered as physical. There is therefore a modification of the effect of the organizational law that was creating aggregates of components without membranes in space, applying its tendency to the production of aggregates by taking elements from the environment. On Earth, the organizational law generates organisms that achieve a unification between the informational envelope and external membranes as well as have plans of reproduction, so that their physical development is structural and systematic, which is the effect of the bifurcation of the law.

This informational envelope allows the first organisms, and even most living organisms, to communicate directly. This is the basis of the sixth sense, which we will develop. We can then put forward the principle of action of the informational envelopes of mobile living organisms. The organizational law commits to the creation of highly mobile living organisms that have an interior, a body membrane and internal membranes for all their organs. The internal organs act to integrate nutrition, ensure movement, regenerate failing cells and ensure reproduction. The most important fact, for these autonomous organisms that live in a space where the action of the Earth's informational envelope is present, is that the membranes of their organs are reifications of the generative informational envelopes and are linked by networks of nerves and veins. The informational envelopes that are highly active for communication with the environment are located in the sensory organs, in the brains and in the reproductive organs. There is indeed a unification between the informational envelopes and the membranes of the organs in mobile living organisms, which is an application of the bifurcation of the creative organizational law of living things.

The creation of sexual reproduction is the implementation of a powerful organizing principle, allowing the very specific development of all living species.

5.2.7. *The creation of sexual reproduction*

For the organizational law, in the immense network of informational links between the multi-cellular elements created, it is a question of giving creative and organizational autonomy to living organisms by making them as autonomous as possible, going much further than the cells that duplicate themselves, allowing the creation of organisms composed of multiple organs. It is thus a matter of producing living organisms with internal reproduction plans and, especially, allowing the evolution of these plans. These plans are therefore dynamic, in order to be able to evolve. Plans are shared between a female genitor, who physically generates the new organism, and a male genitor, who initiates the process and transmits a new part of the plan. The action of the bifurcation of the organizational law is thus to constitute autonomous and very finely constituted organisms with multiple organs, choosing their moment of reproduction and to live in structured groups to carry out these reproductions. There is a bifurcation of the law, both at the level of autonomous organisms with their own reproduction plans and at the level of groups of organisms, to be able to form associations and societies, to reproduce themselves and to ensure the development of the life of the new organisms created. It is therefore a multi-scale bifurcation, which employs a maximum process of reproduction.

The development of life is based on the evolution of sophisticated and rational plans for the reproduction of organisms. Such design plans to produce new organisms are not generated by chance encounters of molecules nor by mere malfunctions, for otherwise there would only be the production of continuous organizational chaos, and not the generation of new stable organisms. There is a general tendency exerted on all organisms in creation, tendencies to produce sets of stable organisms that form species and tendencies to produce new organisms to produce new species, with the functional means to make these productions. The tendency to create new organisms will therefore have specific purposes.

5.2.8. *Reasons for the production of new organisms*

The informational law applied in the production of living things has undergone a bifurcation, in the sense that it will focus on the creation of entities with their own membranes, unified with their informational envelopes. These membranes will allow the realization of the functional system of each organism, defining the structure and organization of its interior and regulating its internal and external communications. They allow the creation of reproduction plans in cellular sets, so as to design new organisms. The organizational bifurcation thus commits to the continuous creation of organisms through the process of reproduction, in order to be able to continue the evolution of the first cellular organisms in an ongoing manner. The always continuous informational communications between the organisms and the general contextual informational envelope is able to engage the execution of the developmental plans, so as to modify themselves to generate original improved organisms, increasing the richness of living things that must not cease to unfold on the planet. There is therefore an informational relationship between the generation plan and the informational set of the context, from the outside. To create a new species, a series of males and females must be generated, forming a first group of individuals of the species that will be able to amplify by similar reproductions, which is something a local generation anomaly in a reproduction cannot do.

We must therefore consider that life was created on Earth by a bifurcation of the organizational law that formed all the material elements in the space of the Universe. The organizational law will engage on the Earth to create elements with physical membranes that reify their informational envelopes, giving them the capacity to generate generational plans for their reproductions. We thus have a global living system all over the planet, which we can call the Gaia system.

In this approach, the definition of a living organism is therefore the following.

5.2.8.1. *An evolved living organism*

An evolved living organism is a complex autonomous system endowed with autonomy, which has a physical body membrane that is unified with its informational envelope, possessing an architecture composed of organs that must maintain its active self by cellular regeneration, which is designed to conserve its energy by either directly consuming photonic energy, like plants, or by eating other organisms, like all mobile organisms. It must also reproduce itself by its strong tendency to do so. These organisms are in direct communication with each other in their geographical domains, using their senses and their limbs to move. They are elements that are part of groups, the whole forming a set that is in a continuous deployment organization.

We can provide a very important property of the life of living organisms, which is of course applied to humans.

5.2.8.2. *The development of energy in the functioning of groups*

Practically all living things live in groups. Being in groups and using their membranes, and their informational envelopes, generates an informational membrane of the synthesis of the group that is used by the substrate, which is sometimes temporary if the group is temporary, but which incites the development of the state of each individual of the group, giving the individual the energy to act in this community. This functioning in groups is maximal in humans, going as far as the formation of civilizations composed of multiple social structures deploying in very structured sets over the whole Earth.

We can now specify the characteristics of the very important concept of sexual reproduction, which allowed the generation of innumerable living species.

5.3. The informational action of reproduction of living things

The reproduction of primitive plants was done by fission, as in the case of green algae. Evolution was achieved by generating sexual reproduction. Two sexual cells, a male and a female cell, merge and this is the beginning of fertilization. We are going to develop the case of living organisms producing sexual reproduction.

In a living organism, the organs that constitute it have communicative physical membranes that are also informational. It is the communications between these membranes that make it possible to constitute the structure of the organism and that will then make it possible to define its state and its behavior.

The informational characteristics of a living organism in its environmental space are:

– the spatial localization and the usual temporal comprehension in its domain in relation to earth time with day and night;

– the quantity of information simultaneously comprehended on the membranes forming its informational envelopes;

– the quantity of information sent to the outside through its informational membranes;

– the evolution of the internal structure–body structure link of the informational interface at the level of energy and information exchange;

– the dependence of the organism with the elements of its environment;

– the degree of proactivity of the organism, taking into account the other organisms in its environment and the importance of its internal regulation and autonomy;

– the energy required and used;

– the tendency to reproduce: the periods of the tendencies to this action are fundamental.

The geographical domains, with their characteristics and the environments of the living organisms, which are composed of many plants and which contain groups of living organisms, all have informational envelopes of transmission. It is these envelopes that will incite the generation of evolutions in the life domains of the organisms.

We must first consider the existence of informational envelopes of environmental spaces. The organizational law can localize its informational action on the Earth in specific environments, which range from small areas to vast expanses. These environments will have informational envelopes that will induce the organisms in them to multiply with certain characteristics that they can share or develop. This will be reproduction's primary tendency: to increase and improve a local species or to generate a new one. We would also put forward that these environmental envelopes will have morphologies defining elements that will influence the informational envelopes of living organisms in the process of reproduction. There is a fundamental rule to the action of the organizational law that concerns its action in the local structures of organisms.

5.3.1. *The fundamental rule of the organizational law that formed living beings*

The numerous molecular organizations leading to life will develop at all levels on Earth and the action of the organizational law, which has had a bifurcation, will be a strong incentive on certain structures, at certain times, to make certain evolutionary effects occur. The organizational law is the system on Earth that generates structured developments and assesses the developments to be continued and expanded. An autopoietic system without any organizational control would only manage probabilistic actions, whereas the living system, conduced by the organizational law, has morphological and cognitive control that operates on all its elements at multiple scales, defining the lines of force in the areas to be developed (Varela 1989). It has a dual level of action in the controlling substrate, that of the representation of the activities of living things and that of the incentive to undertake modifications of living organisms.

The terrestrial space thus forms a very complex system at the physical and informational level, and it is necessary to consider the properties of its innumerable information fields, which constitute its informational space. This informational space, which will immerse innumerable fields, will have both important elements and others elements considered as secondary. There is therefore a morphology that characterizes these informational elements. We will call each important dynamic form of this informational space a morphological pattern. This implies that the informational space has a conformation that has a geometric type, where some components are dominant and others are less important, are linked or are temporarily isolated. We are going to pose that these morphological patterns are the elements that are going to give the important information to the organisms in the process of reproduction. They form the geometric and cognitive informational representations of the informational elements of reproduction, and of the regulatory elements.

We assume the fact that there is always a set of coactive morphological patterns in the envelope of organisms and in those of domains, thus describing the levels of organization in generation in any informational envelope.

5.3.2. *Morphological patterns*

All living organisms have spatial and temporal permanence, as well as the specific environmental domains in which they live, and all have informational envelopes that have morphological characteristics consisting of sets of highly active informational forms that are their morphological patterns, and which make other informational forms less important. These morphological patterns have characteristics of permanence, which produce the major characteristics in the whole duration of activity of the informational envelopes of the organisms. There are patterns that ensure the evolution of the organisms. In the morphologies of the informational envelopes, there are important, dominant and sometimes permanent characteristics that are morphological patterns supervising the others. The dominant morphological patterns impose the characteristics of all communications and make it possible to develop the levels of organization, including, for example, those leading to the specific characteristics of reproductions, giving the physical characteristics of the organs. These dominant characteristics are tendencies towards permanence or transformation. Thus, in all the informational envelopes of the organisms, we have dominant informational forms that impose their characteristics. Let us note that the dominant morphological patterns in the informational envelope of an organism are influenced by the communications with the groups of living organisms in its environment. Dominant morphological patterns have much more informational energy than other patterns. In addition, there is a very strong existential relationship between the organism and its living domain, which constitutes a communication between the morphological patterns of the body and those of the external world. The control carried out by the organizational law on the informational elements of the organisms is thus be carried out by the morphological patterns, which makes it possible to carry out a very precise control, which is both locally and globally sourced.

Morphological patterns are the very active informational forms, whose action will give the characteristic of temporal continuity to living species and their organisms. They are the elements of informational organization that ensure event continuity in living things. They are also informational elements that develop and generate new patterns, thus ensuring the complexity and the evolution of the whole living system through information fields.

We have shown that informational envelopes can be represented by sets of informational agents (Cardon 2022). As any living organism is made up of processes of actions in strong coactivity, its informational envelopes can be represented with a set of morphological patterns characterizing the state of action of these informational envelopes. There are patterns located in the organs, which ensure their relations and active patterns in the general envelope of the organism. This set of morphological patterns thus forms the dynamic image of the functioning of each living organism, which is in complex relation. The key to the problem of evolution is then in the communication between the morphological patterns of the informational envelope of the female progenitor at the beginning of reproduction and those of the environment, including those of the male progenitor.

5.3.3. *The influence of an external morphological pattern on a living organism*

Any living organism is in a strong informational relationship with the domain in which it lives. Certain elements of the environmental domains, including groups of living organisms, will be able to influence it indirectly by participating in the representation of its tendencies in its informational envelopes, such as stability, evolution, selection or disruption. This influence is achieved by sending powerful information in the form of dominant morphological patterns that will be activated in its informational envelope, modifying it. The internal and external morphological patterns constitute the forms that give the activities' major meaning to the envelopes and the membranes during the process of reproduction. This is the tendency to achieve continuity through a similar reproduction, or to achieve it with minor adaptive modifications, or to achieve it with a transformation into an original organism.

You could say that there is a certain randomness in the evolutionary phenomenon. Either the morphological patterns of the informational envelopes coming from the organisms of the environmental domains engage in reproduction, through modification of the plan of development by applying certain characteristics of their organisms represented by their patterns, or there will be a tendency for permanence by the sole action of the dominant internal morphological patterns, therefore using the reproductions plans without modification.

We must pose two hypotheses on the process of action of these morphological patterns, which are external to the female progenitor:

– Prior transmission hypothesis: the major morphological pattern of a set of organisms in the environment influences one of the progenitors, the female or the male, which thus has its reproduction plan modified. This progenitor will generate modified organisms when it participates in a reproduction process. If both progenitors are influenced, then they can no longer generate organisms that are similar to them, only different organisms.

– Transmission hypothesis during the process: the major morphological pattern of organisms in the environment influences the reproductive process while it is taking place. This dominant pattern always remains active and dominant, and influences all the reproductions of the group of organisms, which will thus globally generate a different group of males and females.

With the hypothesis of the prior transmission of information by patterns, a new species is created in the same environment and the old species that no longer gives offspring disappears through natural deaths. It is probable that the hypothesis of transmission of the dominant pattern during the process of generation is the most common because, in this way, mixed groups of new organisms that live in the same environment are rapidly formed, and when these influenced progenitors leave the environment where the dominant morphological pattern is active at the time of the transformation, they can, in another environment, begin again to generate organisms that are similar to them.

In any reproductive process, there is communication between the morphological patterns of the organism's envelope and the morphological patterns of its environment, and these communications are organizational. The envelope of the local ecosystem, composed of the envelopes of the organisms and groups of organisms that are active in it, represents the situation of the progenitor in its field of activity and also represents the importance of the other species, as well as the spatial availabilities that exist in this environment. With these organizational comprehensions, the organizational law will generate appropriate dominant patterns in the envelope of the reproducing organism, by sending strong informational messages. The morphological patterns of the environmental domain of any organism typically form a well-elaborated representation of the characteristics of the environment, as well as the importance of the material and living occupations. For the organizational law in the processes of reproduction, it is a question of defining the corporal adaptations to be carried out and to potentially specify the possibilities of deploying new species, where it is appropriate and feasible. This is how the morphological patterns have influenced the realization of new organisms from the initial ones, developing plant and marine organisms, then all the plant and all the mobile living organisms on the land.

We can see that the organizational law has a major role in the organization of living things which have been deployed wherever possible, in the sea and on land, developing organisms that move on land and even flying organisms. All these mobile organisms have developed to occupy the whole ecosystem, which is generated by plants and the balanced atmosphere containing a lot of oxygen. It is also true that predatory organisms have eliminated weak organisms, as organisms live among one another, eating other organisms or eliminating them to occupy their domain.

We can now mention the formation of brains, allowing autonomous and mobile living beings to understand and plan their usual behaviors in their areas of deployment.

5.3.4. *Brain formation and sensory comprehension*

The senses of comprehension of the elements of the environment, as well as the monitoring and control of the physical behavior of the body, are characteristics interpreted by the production of representations in the brains of living organisms. In their representations that are continuously produced during the awakening phase, the brains generate mental representations making it possible to comprehend the environment, to manage the behaviors and to foresee actions. These productions of sensitive representations are typical generations of the brain, which is an organ of current and planned management of behavioral activities. This complex organ has been created by the creative control of the organizational law, so that all mobile living organisms are endowed with it, thus possessing a center of management of their behavior, giving them considerable behavioral autonomy. During their constitution, which is done in relation to the constitution of the other organs of the body, brains will have morphological patterns that are active in the basic cerebral morphology, which will represent the fundamental tendencies of the organism. In this sense, we can say that the fundamental tendencies of the species in the created organism are indeed inherited. Every brain has a conscious and an unconscious mind that continuously activates the fundamental tendencies of the organism. The case of the brain of *Homo sapiens* is an amplification of the bifurcation of the organizational law, which we will present, endowing it with very specific characteristics in the ensemble of living organisms.

There are thus two types of possible actions in the process of reproduction of an organism: the process of reproduction is located in a specific and limited physical domain. If the generated organism is created with modifications by anomalies in the reproduction process, its purpose will be quite weak and its state will not propagate to a new species; it will instead be isolated.

There are states of reproduction where, during the conception of the development plan of the new organism, a possible transformation is propagated by the influence of morphological patterns. There, at the precise moment of the generation of the development plan, a bifurcation is created in the reproduction process and a significant transformation of the organism with new organs takes place. The condition for this bifurcation to occur is the following, when defining the plan for the creation of the new organism:

– Significant availability of superior-level morphological patterns from the environment of the female progenitor that engage in strong coactivity with the reproductive process of the organism to influence it.

– Creation of an alteration of the morphology of the generation process of the new organism, leading to a modification of certain characteristics and organs.

There is no status of permanence of the dominant morphological patterns that have caused the modifications in the progenitors, there is simply permanence of these patterns in the environment, if it does not change.

There must be temporal continuity of this strong coactivity between the newly generated individuals and the environment, so that there is a continuous series of modifications of the organisms with respect to their initial designers, and that a new species can be generated.

For there to be an alteration in the generative plan of the new organism, there must therefore be temporal concurrency in the open state of the generative plan of an organism in a reproductive state and the availability of morphological patterns in the environment.

5.3.5. *The external organizational attractors hypothesis*

We posit that the informational envelope of a living organism is constituted with characteristics of its morphology that specify the activity of all its elements, as well as the fact that this morphology is, in a certain way, coactive with the one defined by some groups of organisms of its environment. We then put forward that characteristics of this external informational morphology will be organizational attractors, which are formed by morphological patterns remaining dominant during the reproduction processes of a specific series of organisms, thus generating the continuity of the generated species and forming a new set of similar living organisms that form a new species.

An important hypothesis can be made about the functioning of the substrate, with its swarms of morphological patterns and its tendency to be organized by the action of the organizational law. We can posit that there are specific patterns that assess any current situation in its environment, in order to envisage the future situation that will result from it. These are the patterns of assessment. The latter would therefore be the substrate's capacity to foresee the future in order to better organize it, with choices of inhibition of certain unfavorable aggregates.

5.3.6. *The possibility of predicting the future of any situation in progress*

The set of available patterns of the substrate leads to the assumption that there are specific assessment components, as well as morphological patterns of assessment, that give an image of the probable future of any situation engaged by structured elements, so as to be able to organize the continuous evolution of these situations in the world, by inhibiting certain engagements to allow other more stabilizing ones. The notion of time that unfolds on the informational substrate would thus also be a notion of comprehension of the most probable future of each situation in progress.

This existence of morphological patterns of assessment make it possible to regulate the evolution of groups and species well, by being able to inhibit the evolution of certain situations by activating patterns of inhibition. This would mean that the world does not simply move forward in committed organizations but alters the commitment of certain situations to allow for a best possible evolution. We will see that this possibility of having such patterns of assessment of the possible future has a significance in the action of the sixth sense in humans.

5.4. The human species in the organizational evolution of living things

In the evolution of living species, we must specify how human beings who feel, think and conceive were constituted. We will determine how they are positioned today in the ecosystem, radically investing the entire planet.

The development of the neural system is the development of the specific central organ of the organism, which generates representations of the surrounding space with the exercise of its will to specify the themes, making it possible to behave much better in any environment. This gave those who were endowed with it a significant advantage over those who had smaller brains. This advantage was comprehended by the informational networks of the substrate in the groups that were thus favored, and committed to further increasing the development of these brains during the reproductions.

This creation of a particularly developed neural system in *Homo sapiens* would lead them to develop speech using phonic and conceptual memory, which already existed in elementary form in primates who simply used sound effects to inform their companions of certain actions to be taken, which is the case today in bonobo monkeys who are very strongly socialized. This ability to generate mental representations to plan actions committed *Homo sapiens* to a great modification of the functioning of their social groups, and also to develop more and more complex languages. As a result of their very good prehensile capacities, *Homo sapiens* had the ability to manipulate objects and stand upright, which was a new posture among mammals. They constituted societies, and this would lead to a continuous and strong evolution of their aptitudes through the action of new morphological patterns during their reproductions, to ultimately reach *Homo sapiens*, practically 300,000 years ago. There would be evolutionary reproductions by choice of the reproducers, which would be in conformity with the evolution of the social and economic environments, where they created adapted sites. Additionally, the organizational law was strongly active, continuously transforming the newly generated *Homo sapiens*, so that they would be even more efficient for the realization of the elaborate life domains they could build. Lithic tools were found to have been developed by australopithecines 3.3 million years ago, and *Homo sapiens* were very sophisticated tool makers.

5.4.1. *Creation of* Homo sapiens *as a result of very strong evolution*

The creation of the thinking man was only possible through the very important evolution of the hominids, with the progressive creation of *Homo sapiens* who formed very social groups, who could very strongly conceive the characteristics of reality and who was physically capable of transforming a lot of things in their environment, planning and realizing these transformations. The evolution and development of the *Homo sapiens* groups happened by generating bushes and by the fact that they lived in separate domains. In these different contexts, the informational envelopes of the domains would influence their reproduction, so as to give birth to *Homo sapiens* that were increasingly powerful in realizations and in creations. And within these different groups, natural selection would occur. The organizational law realized a new application of the bifurcation to create the *Homo sapiens* genus, which would live generating thoughts that always represent time and duration, with a memory and a very organized ability to imagine, transposing their creative ideas in the form of realizing innumerable physical constructions. For the organizational law, it is a question of generating a system made up of living beings that are capable of generating systems, as it does itself, in its organizational creation of the world. However, does the *Homo sapiens* who thinks understand the globality of life that surrounds them, do they have a sense to generate this type of representation?

The generation of brains with functionalities beyond simply sensitive comprehension is a major application of the organizational law applied to living beings.

5.4.2. *Organizational action of the formation of the human brain*

The case of the considerable expansion of *Homo sapiens* brains in living beings is the result of a new application of the organizational bifurcation, in the formation of living organisms. This bifurcation has made the morphological patterns of development of the central and very powerful characteristic of the *Homo sapiens* brain be dominant in the formation of the individual. Its current development has reached a constructive limit in the application of the organizational law. The human brain is an organ that has a fundamentally organizational behavior with the neuronal elements that compose it, producing complex mental representations by emergence, under voluntary conduct, and using them in very detailed comprehensions that generate sequences of experienced thoughts. The human brain, which memorizes its comprehensions, as well as creates and memorizes languages and innumerable conceptual forms, can generate very finely elaborated complex conceptions by pure creation. This is an application of the organizational law that generated the living being, but which does not plan in a spatio-temporal way, always inciting to improve and reorganize the emergent structures realized by the informational components that form the substrate of the Universe. It is important to remember that the brain is a system with two components: the conscious mind and the unconscious mind. The unconscious mind, which generates fundamental tendencies, some of which are very dark, is not totally controllable in humans, except by learning to control it through education and the exercise of will.

The important development of the *Homo sapiens*' brain will take place with morphological patterns that will focus on the design of this organ, so as to produce a series of human beings that are capable of deep reasoning, of generating a very powerful organized memory, of speaking and of conceiving a lot of things that they can apply to planning and material achievements. Many specific species of the *Homo sapiens* genus will be generated and then disappear in ecosystems where the climate and the context to feed themselves will become very difficult, or where power struggles will lead to extinctions. Today, only one human type is left on Earth, the *Homo sapiens*, whose numbers have grown considerably and who are everywhere, having formed very dynamic societies and civilizations. These developments in humans have been achieved through their ability to make tools with very fine prehensile hands, designing them through the conceptual work of their brain, with their ability to generate elaborate representations relating to the structure of things in the world in space and time, to conceive and carry out structured planning of constructions, to generate spoken and written languages, and their social ability to constitute tribes, followed by very strongly structured and hierarchical societies, going as far as civilizations. Humans have also reproduce a lot by dominating all the other animal species, traveling all over the planet and spreading out.

The notion of groups of living beings is significant, as is that of aggregate in the constitution of the Universe. The organizational law always operates at the level of groups of individuals, by assessing the informational characteristics that they produce in their actions. For *Homo sapiens* always living in groups and structuring these groups by their collective actions in their environments, the assessment of the organizational law will involve the dominant informational characteristics that have led to the realization of the actions of the groups.

5.4.3. *The importance of informational links between groups of humans*

Human brains generate representations that are sequences of developed characteristics that are quite specific. There are groups of reflections, where they often all think very strongly together about similarly themed characteristics, thus defining a strong local morphological pattern that will be comprehended by the informational context of the group, to be memorized morphologically as an informational element. It is such patterns that are able to introduce modifications in the group's regulatory and deployment actions. Therefore, the way in which group members think together is an informational action that will also have organizational consequences for the future of the group, through specific modifications in reproductions. This is a particular sensitive comprehension, where each individual integrates the sensitive comprehensions of the others, making it appear as a sixth-sense characteristic.

We have laid out the notion of groups to generate new morphological patterns, because such generations are organizational processes of groups and not local processes of the behavior of an isolated individual. The organization of the living being is absolutely collective, comprehended and appreciated by groups that have coherence, and it is therefore the actions of the groups that make the representatives of the living being evolve, by generating more adapted and more efficient individuals.

We can come back to the advantage of being in groups among living beings and conclude that humans are absolutely social individuals, who can only exist as a human if they participate in well-organized and regular social and cultural meetings.

5.4.4. *Power of group participation in humans*

Human beings have a strong capacity to think and, therefore, to behave by thinking. However, humans must live in groups to acquire their knowledge and their cultures, as well as to generate collective mental forms that make it possible to deploy an informational envelope of the groups, which reinforces the energy and each member. The functioning of humans in well-structured groups, where each one can understand what the other feels, has allowed them to generate societies. Let us note the fact that a human participating in a specific meeting in a well-organized group will give them informational and mental energy, and will allow them to acquire deep values, which are values of sharing. This is the reason why religions and all social and cultural groups were created.

We can say that human beings, who never cease to develop their culture of conception on reality and their collective life, must organize regular meetings in very strongly coherent groups. In these groups, bringing together humans who seek to collectively deepen problems by sharing the same values, the envelope of informational synthesis becomes very strong, very organizational, and allows the development of strong sharing comprehensions in minds, and thus in memories. Human societies greatly developed these structures of organized groups over the long-term, until the invasion of computer communications enabled the transmission of messages, images, documents and films. Human beings are then alone in front of their screens, no longer feeling the presence of others with whom they can communicate: there is a closing in on oneself with virtual sociality.

Human beings have a very particular psychic system that gives them very powerful conceptual representations with − what Sigmund Freud defined very well − two very great fundamental drives that orient all the others: the life drive and the death drive (Freud 1966). With the life drive, humans have the natural ability to conceive, to represent themselves, to give and to share with all others in their environment. With the death drive, they have an aptitude to strongly dominate their exterior, to dominate the other by going as far as to consider that they must eliminate the other. These two impulses exist in the psychic system of each human, they cannot be eliminated and lead to the formation of their tendencies and intentions, to characterize their mental productions, and therefore structure their societies. It is thus necessary for them, both socially and culturally, to intensely develop the life drive to constantly reduce their death drive, to commit himself to build societies going in the right peaceful direction and to make fundamentally convivial humans.

Today, it can be said that a human is somehow able to directly communicate with another human in an immediate way. This possibility of direct communication is due to the discovery of mirror neurons and their interpretation in 1991 (Gallese and Massimo 2015). These mirror neurons, which are also present in many animals such as bats and rats, are motor neurons that fix the purpose and meaning of a mental representation that is going to be generated and comprehended, and comes from the representation generated by another living being that is strongly considered. In a human, the mirror neurons have the potential ability to directly produce a certain representation in the brain of what the other human, who is very close to them, is thinking or feeling, without proceeding to complex analysis and reasoning to determine what the other is really thinking and feeling, according to their attitude and the movements of their body. This is a situation where a human can think and feel something, and another human who is interested in them, who observes them, can see a representation that is directly generated in their brain that has similarities with the representation generated by the one they are observing. There is therefore a process of direct communication between humans, which is considered to be the basis of empathy, that is, of the sensitive and emotional sharing of one with the other.

We can therefore say that the informational envelopes of humans and their brains can enter into communication to exchange particular informational elements belonging to the generated mental representations, so as to carry out a certain type of informational transfer. This is a very strong characteristic in group life.

We can acknowledge that certain animals have a comprehension of reality that makes strong use of the informational system, which would be specific to them and would allow them to comprehend an inaudible or invisible presence by the eyes, as well as perceive the extent and state of their environment with those who are comprehended to be in it. With this sense, they can also move over very large distances, with great accuracy of the trajectories taken, perfectly locating their point of departure and return, as is the case for bees and migratory birds. We speak, in this case, of the sixth sense of animals. Human beings naturally only have five senses and do not directly perceive the informational characteristics of the terrestrial reality. They do not have the comprehension of presences outside of seeing them with their eyes and hearing them with their comprehension of sounds. They live within themselves a lot, developing very powerful sequences of concepts and linguistic reasoning, which has allowed them to go further than the life of sapiens in simple groups, to found their civilizations with today's considerable technological developments. We will develop this sixth sense in the rest of this work. We will first define the characteristics of the functioning of the human brain.

6

The Interpretation of Neuronal Aggregates

6.1. Introduction

The neuronal system operates at the level of the parallel production of multiple neuronal signals which, by their associations and aggregations, form a very complex whole that can be interpreted as a structure of dynamic forms that combine; a structure constituted by activities and informational exchanges, which carry sensitive and cognitive indications at a certain level.

A thought is formed of many significant characteristics about what is comprehended, some characteristics being important and others being secondary, contextual, associated or even opposed. The number of these characteristics is significant, but does remain finite. We assume that these characteristics are represented by the action of significant groups of neurons, which we will call significant neuronal aggregates, and which are active among one another through energetic and informational communication. These groups can be interpreted as the dynamic forms that contain information to generate the significant characteristics of thoughts. These neural aggregates are active among one another through being solicited to establish relationships. They are then activated on larger scales, so as to form aggregates of aggregates, which will be the form of the expressed emergent thought.

6.1.1. *Form of a thought*

Here are the characteristics of the form of thought:

– It is a complex and essentially dynamic element, made up of the energetic and informational movements of neuronal aggregates that unfold and activate simultaneously on several scales.

– It is a mainly dynamic construct, using the memorization of the characteristics of some forms that have already been realized.

– It is a dynamic construct that expresses itself, that is used by the system that produces it so that it can be experienced, and that only lasts the ephemeral time of this conformation, in order to be able to continue with other generations of forms that will be the following thoughts, in a continuous process in the awakening phase.

– It is thus a process that shares a certain similarity with the informational generation realized in the Universe, but on a very short time scale.

Each thought is thus a construction realized in a sequence of realizations of thoughts, with strong reconstructions using forms that have been expressed and that are memorized. For the most part, the reconstruction of forms is potentially memorized in more or less similar forms, using a memory of conformations and not a memory of elements, and this at each ideal generation.

A thought is a mental representation of something specific, which is constructed, felt and appreciated to be used and reused, and which systematically engages to produce others.

6.1.2. *Constructivist definition of the notion of mental representation*

In the neuronal network, a felt mental representation is the spatial, energetic and above all informational generation of the conformation of an organized flow of numerous neuronal aggregates, which constitute a precise and stabilized structure for a brief moment. This flow will appear in the form of an organization of neuronal aggregates constituting a spatial conformation, formed at the minimal level by strongly connected neurons that are made up of aggregates, which are themselves connected. It is a dynamic organization, sensitively appreciated by the system when it is constructed and available to be comprehended. This representation will be comprehended through exploration by elements representing organizational forces, exerted by multi-scale actions on the physical and/or informational components that constitute it. We could say that it applies to things of the world by some of its morphological characteristics, which will always be linked to the type of thing comprehended. It represents, designates and expresses a real or abstract thing, by its aspects and its characteristics, and always at multiple scales. This representation, which is a representational construct, is made to be transformed into another more or less different one, which will be the next representation, so as to constitute the flow of generated representations forming thoughts and which is impossible to interrupt. A representation is therefore not a simple functional state but a dynamic organization of informational forms, which is constructed in a continuous sequence and then altered during its comprehension by the conscious mind (Marchais and Cardon 2010).

We specify the process that produces the representations by showing the active and organizational elements that it contains:

1) The very many elements based on neurons and synapses, which constitute the basic elements, whose function is to activate and connect.

2) Actions aggregating these basic elements to constitute spatial, energetic and informational forms, of a certain typicality.

3) An organizational action aggregating all these forms in a continuous way, according to different energetic and spatial scales, in order to constitute the form of the representation. Any representation on a certain subject has a particular spatial and energetic conformation to the expressed subject, and this dynamic conformation can thus be described geometrically by a constructive and mobile movement with energy, conveying a lot of information on different scales.

4) An action of certain basic forces exerted on the construction of this representation, in order to bring out its global meaning with certain important characteristics, mainly when it is being organized in a coherent way, which is the act of consciousness that experiences its representation. There is, therefore, a notion of a morphological pattern that can be

comprehended. A fact of awareness of something is a dynamic comprehension that expresses a multi-faceted meaning when the representation becomes well comprehended, and the psychic system must be seen as an essentially sensitive and controllable system by itself, and which is centralizing for the perception of things by humans.

5) A general process that leads to using or abolishing this representation that has been activated to generate the next one, which is the process of continuous activity.

NOTE.– Let us note that the process of production of the representation, which is dynamic and mobile at the energetic level, influences the state of the representation itself, by the fact that it is constantly more or less modified as a result of being constructed with relations between its elements and by the intended purpose, and this property provides this construction process with very particular characteristics. These characteristics are indications for the numerous movements that are realized and that are always valued on several scales.

The approach consists of considering this production in a geometrical, dynamic and cognitive way, which makes it possible to give it its characteristics in a measurable domain.

The system generating thoughts in the form of representations will have to be endowed with intentions to produce them, in order to be able to produce numerous types under multiple characteristics. We follow the Freudian architectural model, and we will interpret it and assume that its general architecture is the following, with four instances of the thought system: an emotional processing center, an unconscious mind incorporating an organizational memory, a preconscious mind and a conscious mind. More precisely, these instances, which will be dynamic sub-systems, are the following:

– The sensitive and emotional processing center, treating the five usual senses in parallel and generating the different types of emotions as immediate responses, interpreting the activity of the body, as well as the external and internal data of the senses. Its role is that of the thalamus and the limbic system. This component is very finely linked to the functional elements of corporeality but, above all, to the preconscious and the unconscious mind, with which it communicates constantly to introduce the characteristics of the senses and the emotions into the representations, often generating representations that are essentially sensitive, immediate and reactive, that is, very little conceptualized. It manages the formation and development of emotions in the preconscious mind and participates in their transformation into experienced emotions, in other words, into feelings, by being linked to the conscious mind.

– The unconscious mind and the organizational memory, which localizes the lived experience and can be considered to localize a memory by generation that will be organizational, and which also localizes the impulses. In this memory, which is potential and forms a substrate everywhere available, elements are exhibited by structural and energetic generation, with memorized events being in the form of structures of strongly connected basic elements. This component thus reifies the structures and the very dynamic organization of the lived events, by representing an organizational memory where the memorized elements are potential forms in strong relations, which can be activated and again become the comprehended memorized thing.

– The preconscious mind, where active and structured elements from the unconscious mind and the emotional center pass through, with elements of control shared with the unconscious mind, so as to form significant active aggregates for the representation that is being formed. The preconscious mind will therefore contain patterns controlling the generation of informational forms. This component constructs the forms of competing pre-emergent representations, by strongly using a morphological analysis of the active components. This will be the site of a control exercised by the controllers of rational analysis, judgments, postures, desires, sensations and feelings, which will form multiple classes of morphological patterns.

– The conscious mind, where a form distinguished from the preconscious mind will emerge, which will be manipulated to be experienced and perceived according to a particular autonomous process, in a specific sub-system of the meta-level, before being memorized temporally.

Let us note that what will generate and allow us to experience a thought is an emergence constituted by a complex dynamic organization of neuronal aggregates, constituting multiple specific spatial forms. We will therefore say that there is a form of thought that is experienced in the conscious mind, as well as the physical emergence that founds this feeling, which is distributed in the four instances of the psychic system.

6.2. The systemic layer and the regulators including the informational regulator

We assume that thought is the dynamic production of a psychic system that does not cease to construct and reconstruct forms constituted by aggregates, by using its great possibilities of neuronal actions. We can also assume that there is a fifth architectural instance in the psychic system defined by S. Freud, after the components of emotion, unconsciousness, pre-consciousness and consciousness, a component that is essentially an activator, and that realizes the synchronized control of the activities producing the facts of consciousness that are comprehended. This fifth instance is called the systemic layer of the psychic system. It has the characteristics described in the following sections.

6.2.1. Systemic layer of the psychic system

The systemic layer is the dynamic component of the psychic system that makes the production of the thoughts take place by exerting an incentive control on all the elements of the system, at all the scales, including the other instances, in order to put them in coherent activities and to provide the conscious mind with a representation to value. It allows the effective realization of emotion as a movement of training and domination to be established between the unconscious, the preconscious and the conscious minds, taking into account the center of emotions. This layer with continuous functioning is the major process of setting all the components of the psychic system in action and in co-activity. It allows the reification of the Ego of the system, with its continuous relation with the external world through the component of emotions, expressing the posture of the organism in front of reality and in front of itself.

The systemic layer functions in a continuous way in the human psychic system, except if humans put themselves in a state of openness at the informational level alone of the external world. It is the system that allows humans to be aware of the notion of the passing of time, and therefore to be aware of death as a rupture. It will be essentially composed of energetic and informational control elements, which will be specific morphological patterns, operating in co-activity at multiple levels, allowing the generation of organizations of neuronal aggregates representing the physical, dynamic, energetic and informational form of the thought, and which emerge as a stabilized form for a moment, to be experienced. These elements of control are called regulators, and we will develop their characteristics, their categories and their relations.

6.2.2. Regulators

A regulator is a controller operating on the basic elements and aggregates of basic elements of the psychic system, so as to activate them and make them organize themselves. It consists of specific morphological patterns carrying out the activity of informational control flows. Some of these regulators are specific and operate on other regulators, in order to realize a multi-scale control which we will call organizational regulators. A regulator thus operates as an informational and electromagnetic line of force that can be deployed in a loop, and that operates on the elements that it must control. It is the main control element of the system and there are regulators on many scales, forming a very dynamic regulatory space. There are regulators operating morphologically to drive tendencies and generate representations.

The regulators form layers of links between the aggregates to allow them to activate and co-activate each other. These layers, which we represent by patterns, are basically composed of synapses and chemical elements using energy and information. They are the elements of a network that meshes the whole system, that connects all the neuronal aggregates and that even operates within the aggregates by ensuring their autonomy. There are regulators to conform the aggregates of the representation into the right conformations according to the themes, and there are a whole set of regulators allowing the system to function. There are organizational regulators controlling the regulators. There are therefore regulators determining the fundamental drives, the tendencies, the emotions, and there are regulators defining the current aim, which commits to producing each representation. There is also a meta-organizational regulator, strongly linked to all the others, which commits to voluntarily produce representations rather than to undergo things in a neutral way, without any intention.

6.2.3. The voluntary choice of the operated aim in the psychic system

It is possible to choose a focus for producing any representation, by selecting a specific organizational regulator. There are regulators for each of the senses, such as the auditory or visual sense, as well as specific organizational regulators for selecting a memory or a theme based on a word, which can be selected from well-structured memories. This choice means that the psychic system can be controlled at the level of its developmental organizational characteristics by the will exerted on it, through the organizational regulators.

These regulators also form a Web of relationships that is not at all independent of the aggregates. There is a relationship of co-activity between the aggregates that have some behavioral autonomy and the regulators that cause them to activate in certain ways, as the system is deeply organizational, unified and absolutely not composed of different functional levels. This linking network is organizational on several scales, because there are layers at the simple level of groups of aggregates and layers at the level of organizations of groups of aggregates, localized in specific areas of the brain. All the organizational complexity is measured in these nested layers, as well as in their very co-active relations with the aggregates.

6.2.4. *The aggregate–regulator rule of co-activity*

In the psychic system, the elements representing the elementary characteristics of meaning (the aggregates) and the elements representing the incentives to control the organization (the regulators), are coactive and dependent. There is a systemic link between these two types of elements, which makes the developments and the expressions of the system powerful, as well as show its fragility, which can be the cause of memory disorders.

The great strength and weaknesses of the human psychic system will be in the coordinated or contradictory activity of the regulators, which has allowed us to specify the origin of many mental disorders in the human psychic system (Marchais and Cardon 2015).

The general diagram of the architecture of the psychic system, with its systemic layer of self-control, is shown in Figure 6.1.

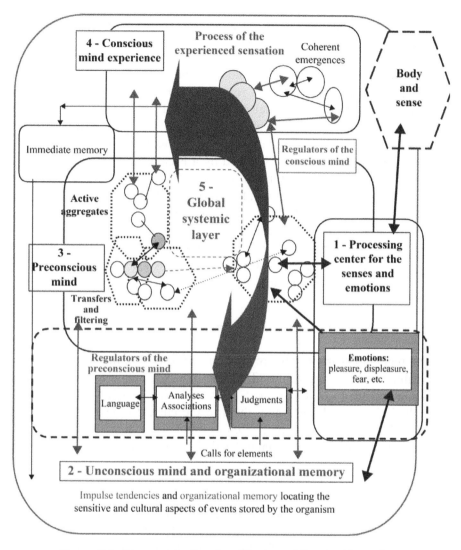

Figure 6.1. *The organizational architecture of the psychic system with its five instances. For a color version of this figure, see www.iste.co.uk/cardon/generation.zip*

This architecture requires detailed specification of the domain of the regulators and the necessary characteristics of the emergent form, in order to be able to represent thoughts that can be tested. We will first specify what the regulators are and what they do, as well as how they operate together in co-activity.

On what basis is the control of the aggregations and of the whole neural organization done to produce the representation that will be experienced? The regulators are the rational controllers of the organization of the co-activity of the neuronal aggregates, representing the normal psychic tendencies and the cultural capacities of the system, that is, both the drives and the tendencies like those exercised by a Freudian Superego on the Ego. There are therefore regulators at the level of the preconscious mind, anchoring the effects of the drives, the abilities to identify and distinguish, to analyze, to reason, to abstract and categorize, to feel sensitively, and regulators at the level of the conscious mind, which reify questioning, conceptualization, openness, self-consciousness, ethical judgments and values. The types of regulators and their relations are obviously the key to the psychic abilities of the system.

6.2.5. *Morphological role of regulators*

The regulators carry out the regulation and control of neuronal aggregates or organized sets of aggregates. They facilitate the active insertion of these elements and their co-activity and, above all, they modify the tone of the controlled elements, by adapting them to the current tone of the conscious mind, and sometimes of the preconscious mind. They thus give the cognitive and sensitive quality of any comprehended emergent representation, as well as its value.

The regulators of the preconscious mind are, in fact, elements shared between the unconscious and preconscious minds. They are dynamic and have a complex structure with networks of patterns, operating at the morphological–semantic level. In the systemic layer, they allow the extraction of the right groups of elements from the memory of the unconscious mind, which are adapted to the aim and to the tendency of expression that is valid in the conscious mind at a given moment. These elements are highly organized among themselves and form the network of possible tendencies which, at all times, give the theme, the aspects and the characteristics of aspects to the perceived form, which are constituted in the preconscious mind and emerge in the conscious mind, allowing it to be inserted into a continuous flow of coherent emergences that are valid for a series of generated thoughts.

The main regulator of the emotional processing center manages the emotional flows that are established in the preconscious mind, which come from the corporeality. It is a highly structured regulator, composed of numerous sub-regulators that correspond to the various types of emotions that can be felt. It produces an emotional layer, a morphological landscape of co-active aggregates in the preconscious mind. This very dynamic layer represents the emotions that are activated in the form of attractive centers, which are transformed, confronted and linked to forms of rational thought.

The regulators of the conscious mind are directly linked to the systemic layer and enable it to be activated and controlled, in order to impose the tone of the preconscious mind on the regulators, and therefore specify the dominant characteristic of the emerging forms produced in the conscious mind. The aim of what will be given to think will be produced by the co-activity of these regulators. The focal point of current emergence, with its related characteristics and complexity, is thus driven by the regulators of the conscious mind. Moreover, these regulators make judgments about the value and adequacy of what emerges with the reality of the environment. These regulators form a network that expresses the Ego of the system in its action to provoke ideal emergences, felt in a continuous way. This network is obviously linked to the network of regulators of the preconscious mind, in a normally directive way, and to the structural elements of the systemic layer.

All the regulators are structured with sub-regulators that specify them. They form networks with quite numerous and relatively evolving local hierarchies, since they correspond to the categorization of everything that can be thought, reasoned, appreciated, judged, felt and wanted by the system. All these characteristics have been strongly developed by psychologists and philosophers.

These regulators each have the following:

– A specific thematic or categorical domain, which is reduced or significant, wide or narrow. This domain of action is defined from the categorization of thoughts that the psychic system is able to define.

– A history, defining them as native to the generation of the brain or created by learning during education or even during operation.

– A co-activity, which is used to establish privileged relations with some other regulators or antinomy with some others, which is able to evolve in time, according to the ideal generations.

– A general scope, which will be the morphological deployment of their co-active actions with effects on the sets of neuronal aggregates, as well as on the other regulators. This activity constitutes the morphological–semantic space of the regulators, which have characteristics expressing the possible and qualified tendencies of the psychic system, and which more accurately define the Ego.

There is, therefore, an organization of regulators. This organization is a spatialized whole, with elements that are informational, energetic and active in the current state, and others that are inactive. This organization is formed in the living thing at the formation of the psychic system, by a meta-regulator that has a unique role: it is the regulator that represents the life drive described by Sigmund Freud.

During the construction of the psychic system, this meta-regulator engages the construction around the regulation, which is the central point, by generating the systemic Web, and places the specific regulators that are created in spatial organization in spherical or pyramidal geometric form, so that some regulators are always dominant and active. There is thus a fundamental drive in the system, represented by a regulator that can take a dominant position (see section 6.2.5.1).

6.2.5.1. *The fundamental drive and the willpower regulator*

The fundamental drive of the psychic system is represented by an organizational regulator operating on all the regulators. It is the fundamental tendency of the system and it is represented by a meta-regulator that we call the willpower regulator. It places the system in a situation of tension in order to pursue its capacity to exist, so that it produces actions and intentionally generates forms of thought.

It is a drive of constructive openness, pushing it to understand, assess, question and communicate. This drive can be reduced, set back for a moment, allowing the meditation posture to appear. If it is permanently reduced, then the system falls into anguish. This fundamental drive produces variations in tendencies and in specific drives, according to the profile of the system, which are dynamic incentive elements of production of the different types of ideal emergences.

The process of generating a representation after the production of a tendency or a desire is given in the following example:

– normal activation of all instances of the system;

– action of a tendency regulator expressing a specific desire;

– action of the willpower regulator and generation of an aim, indicating the theme of the desire by the control system;

– immediate generation in the preconscious mind of a short form of representation, a pre-representation expressing the theme;

– acquisition of the pre-representation in the conscious mind in its form, without modification;

– emergent generation of the representation with action of all the active patterns;

– sensitive comprehension of the emerging representation;

– production of the analysis of the representation and first simple memorization;

– generation of a short sequence of representations, specifying the way to go about reviewing what has been targeted;

– possibly deep memory production of the experienced representation;

– physical action of the organism corresponding to the comprehension of the representation;

– generation of short representations on the memorial facts of the activities that have already been carried out, with the thing comprehended during the physical action.

Figure 6.2 depicts the instances of the conscious mind, showing the representation that will be felt.

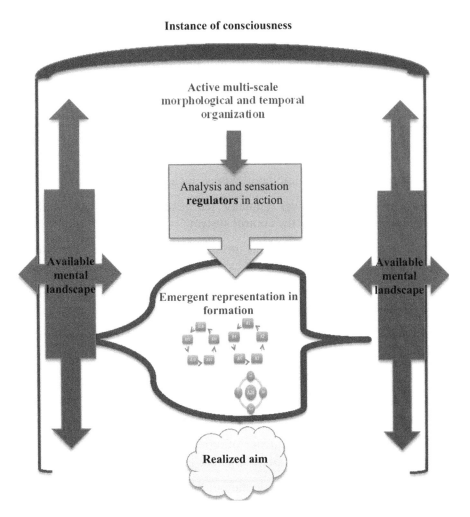

Figure 6.2. *The comprehension of the emerging representation in the conscious mind. For a color version of this figure, see www.iste.co.uk/cardon/generation.zip*

We can present the general process of generation of a representation that will be produced and experienced by the human psychic system.

6.2.6. *Process of intentionally producing a representation on a desired theme*

6.2.6.1. *Beginning*

– Continuous and regular activation of the internal and external patterns of the organism.

– Continuous activation of the sensors reifying the sensitive comprehensions of the organism's physicality.

– Activation of the emotional processing center with the action of its five senses.

– Continuous activation of the organizational layer.

– Activation of the regulators representing the fundamental tendencies of the psychic system.

– Activation of the great willpower regulator to generate a representation.

– Preliminary construction of the initial available mental landscape:

   - activation of all regulators in all instances;

   - activation of the organizational layer;

   - activity of design regulators solicited by the active regulators in the preconscious mind;

   - precision of the current mental landscape according to the immediate memory and the organizational layer.

– Prior activation of organizational memory:

   - activation of the memory layer through the action of the active regulators in the instances and in the mental landscape;

   - sending of some aggregates of memory design regulators that are coherent with the mental landscape in generation in the preconscious mind, by the memory regulators.

– Initial activation of the conscious mind and the organizational layer:

   - activation of the production regulators of the aim in the conscious mind, in coherence with the state of the organizational layer, taking into account information from the immediate memory, the mental landscape and the active regulators;

- action of the intention to generate an aim, by soliciting the action of the aim regulators on one of its possible themes at that moment;

- co-active activity of all the aim regulators to generate a hegemonic aim regulator;

- very rapid release of an aim regulator on an indicator that is activating in hegemonic mode, and launching its indicator theme on the other regulators and on the organizational layer;

- modification of the mental landscape by the action of the regulators of all instances that are in conformity with the targeted theme, and by the action of the organizational layer.

– Activation of the preconscious mind:

- activity and linking of design regulators coming from the organizational memory, by the regulators following the aim;

- formation of aggregates of design regulators by their strongly co-active tendency;

- strong activation of the rational regulators and the emotion regulators on the aggregates, and formation of conformations for pre-representation.

– Activation of the conscious mind:

- managing of a conformation released in the preconscious mind, corresponding to the incentive actions of the active regulators;

- formation of the emergent representation in its final form of design regulator conformations, consisting of aggregates driven by the active regulators of the conscious mind;

- activation of the process of sensation to think about the emergent conformation, by analyzing it through manipulation of its morphological aspects, with the sensation of understanding, pleasure or displeasure. Dominant sensation in the system to comprehend the emergent form, producing the construction of a corresponding synthetic form to memorize it.

– Final action of the conscious mind and the organizational layer:

- immediate memorization of the constructed synthesis of the felt representation, for future memorization in the organizational memory.

– Modification of the preconscious state:

- action of new regulators of the emotion processing center through the sensors, and the return of perceived actions of the effectors in the preconscious mind;

- action of some regulators on the synthetic form that has just been lodged in the immediate memory;

- according to the importance of this synthetic form, the placement of this memory in the organizational memory with local modification by the memory regulators, or the form is left in the immediate memory for use in the production of the following representations, its theme being preserved as dominant in the organizational memory.

6.2.6.2. *End*

This clearly shows the complex internal process that leads to the production of a desired ideal representation, knowing that the neuronal system does not stop producing them when it is in a state of activity and not in sleep. This process is internal and is based on the regulators and the morphological patterns that are active at the moment of production of the ideal form that is going to be experienced. These elements are very numerous and characterize the psychic system of the organism that produces its thoughts. There is therefore a very dense set of informational actions between the aggregates that have informational envelopes during their activity. The big question is, Does this system, which is based on the energetic and informational communications of its aggregates, have direct communications with patterns that are external to the organism, or is it really isolated with a closed functional mode?

Thus, the problem now is to define the communication of this organism that generates ideal forms with the informational world. For there is indeed an informational substrate linking all living organisms and it is necessary to see how to comprehend information from it. We therefore assume that there is a specific organization regulator, operating at the level of the specification action of the psychic system; the organizational immersion regulator in an informational state.

6.2.7. *The organizational immersion regulator in an informational state*

When solicited by an act of will, this regulator, known as the informational regulator, inhibits the action of many other organizational regulators, such as those of reasoning and sensitivity, and opens the psychic system to direct communication with external regulators. It achieves a change in the comprehension and functioning of the psychic system. It can be activated when an aim has been defined and realizes the will to put oneself in a state of informational openness, in order to comprehend the targeted thing at its informational level. It is this informational regulator that cuts off the rational and sensitive functioning of the psychic system for a short time, through the establishment of an informational link, by comprehending informational forms on informational envelopes and then returning to the usual state of the psychic system, so as to cognitively and sensitively comprehend the information received.

It is this informational regulator that allows living beings to use their sixth sense. It is a regulator that can be considered as primary, that is, it exists in all living beings. It allows the activation of the informational envelope of the brain, the latter linking its physical envelope to its informational envelope. It will very often be put into inaction by the human psychic system, which has a complex and permanent internal organization, according to its education and its culture, engaging swarms of internal regulators controlled by organizational regulators to feel and reason using language.

We will now develop the history of computing and AI, which goes all the way to fully autonomous and fully communicating systems.

The Sense of Informational Comprehension of Living Organisms: The Sixth Sense

7.1. Introduction

The generation of thoughts in human beings depends on their current state and their situation in their social context. If they are in an environment where problems have disrupted this environment, or if they are in a group where people are very gloomy without them knowing it a priori, their thoughts will systematically become darker. They are therefore constantly in a situation of sensitive comprehension of their environment. The system, with its active informational links and its dominant morphological patterns, has communicated with the major patterns of its psychic system and has been able to influence them, by being open to this comprehension. This involves the human being's openness to the informational comprehension of things in the world, as well as their tendency towards closure that can cause them to close in on themselves.

We usually attribute a sixth sense to animals (Bomsel 2006). This sixth sense is not like the five usual senses, because it is a global and organizational comprehension of elements that are outside the comprehension of the five usual senses. We will show how this sense is an comprehension of the elements of the informational network that makes the organization of all the elements on the Earth, by developing what this informational comprehension of the communicating informational envelopes is, which is totally different from a visual comprehension of a world of objects with measurable positions in space. We will also explain the notion of magnetism used by healers.

7.2. The five usual senses and the use of the informational substrate

The five senses in animals and humans are systems for comprehending things in space-time. Any sensitive perception gives information about the spatial origin of the object on which the senses focus in relation to the receiver, defining the line of distance. This will not be the case for the sixth sense, because the informational substrate is a virtual space that is comprehended by the informational links, constituting a hyper-network.

The five common senses of animals and humans are as follows:

1) Sight: seeing space with both eyes gives a three-dimensional comprehension of the domain which is being focused on, making it possible to locate objects in their respective positions, to grasp their colors and their sizes, and to follow objects that move. This fundamental sense is used in a continuous way to conceive and regulate the activity of displacement of living beings, so that they behave in their comprehended space of life, to act on it. It is a processing of the photon fields from the observed objects.

2) Hearing: this sense allows us to distinguish the frequencies emitted by the vibrations of elements, as well as to choose ranges of sounds by selecting sounds, identifying their origin and location. The origin of the auditory sense is based on the fields of vibrations generated by our atmosphere. It is a processing of the vibratory fields of the atmosphere.

3) Smell: this sense is highly developed in certain animals, who can smell the presence of other animals and physical things, and thus identify them. It is also developed in humans, who mainly recognize generic smells, perfumes, as well as good and bad smells circulating in the environment. It allows us to identify molecular elements that circulate by the propagation of clouds of molecules, which are structured and then comprehended. It is a processing of the flows of molecules that enter through the nose.

4) Touch: this is the comprehension of the temperature and the general morphology of what the fingers can touch. The origin is the physical and sensitive contact at the temperature level, experienced by the skin of the hand. It is the perception of heat and magnetic fields that the hands touch.

5) Taste: this allows us to appreciate what is going to be eaten or drunk and is very important in terms of the experience of the pleasure of eating, which is present in all animals and especially in humans, who have

developed a very powerful culture of eating in all their civilizations. The origin is both molecular and thermal. It is the processing of heat and molecules that are absorbed on the tongue in the mouth.

These five senses in humans are activated for their comprehensions by specific regulators, which allow the sensitive selection and interpretation of their information in the neuronal system. This information is then experienced and conceptualized using the power of memory and language, so as to establish designations and generate cause and effect. There is an incessant functioning of the human conceptual psychic system that nominally and conceptually identifies the forms comprehended by these five senses.

It is by using these five senses that the development of human societies, and then of the structures of its civilizations, took place, by using the two fundamental drives of human beings: the life drive, which opens to all others in order to accept them, and the death drive, which is the closing in on oneself in order to dominate all others (Freud 1966). However, there is another sense that human beings do not seem to use very well, which is the comprehension and the use of informational substrate to communicate; the substrate that made the Earth and all living things.

7.2.1. *The case of animals*

We must consider the ability that many animals have to comprehend situations that concern them emotionally, without using the usual senses of sight, hearing and smell (de Wailly 2009). There are examples of dogs that have traveled hundreds or even thousands of kilometers to get back to their owner, going in the right direction towards them, without prior knowledge of the right way to reach them. Dogs and cats are apprehensive about the arrival of their master, who is driving by car and is not yet nearby, because they are eager for them to arrive: they perceive their arrival. When animals are in a vehicle, they feel they are arriving in a specific place that they know well, even though they cannot see it, because they are still far away. This is the case of a dog in a car that knows it is going to arrive home, even though it is still a few kilometers away, making it get up in anticipation of exiting the car. In the case of a cattle, sheep or horse farm, these farm animals can detect the arrival of the farmer who is going to bring them food, even when they are not very close and not in view, but on their way. Furthermore, migratory birds have the ability to travel considerable distances to find their nests in very specific locations that they used the year before, and whose route and location they know perfectly.

How do all these animals have this perception of phenomena in the reality of the space-time of our planet, without comprehending them by the five usual senses of sight, hearing, smell, touch or taste, even though they have good awareness of it, while we humans do not perceive these things?

7.2.2. The case of plants

Plants are usually considered to be at the very bottom of the living scale. Plants are what allowed animals to develop by providing viable homes, managing moisture and producing oxygen. We also know that plants, including trees, communicate with each other. They send information to each other in the case of aggression by animals eating the foliage or investing the plant, like insects and birds do. In tropical areas, trees that are on the periphery of large groves generate thorns, so as to allow tree life in the grove. According to scientists, they communicate by sending out electric fields when certain trees are attacked by leaf eaters. Some plants exist that are considered carnivorous, such as the dionea, which has a jaw that allows it to pick up insects to prevent them from degrading it and to feed on them. The internal organization of these sensitive plants, such as *Mimosa pudica*, and those that can open and close their plant jaws, such as the dionea, are worth studying (Hiernaux 2020).

We will put forward a hypothesis on the communication of plants, considering that plants are among the first living organisms on Earth and that they come directly from the communicating informational organization that generated life.

7.2.2.1. The continuous informational transmission of plants

It can be hypothesized that plants, which were among the first living organisms, have an ability to generate informational fields, by their organization and by the informational envelope they possess, indicating their positive or negative states to other nearby plants. This continuous transmission of elementary information would be the origin of the sixth sense in living beings. The latter is the ability to conform to the informational status of everything that exists in a structured and autonomous way on Earth, and is absolutely consistent with the informational functioning of the organizational law. This sixth sense would then be developed as a special organ in the newly generated autonomous animals and would combine with the other senses they possess.

This hypothesis assumes that the informational Universe has directly transposed an ability to use its informational envelopes into the initial living being, so as to communicate directly as all structured elements do to organize themselves. It is the use of their informational envelope that unifies the branches, the foliage and the roots. In animal species that are endowed with mobility that will then be generated by evolution, this will be a sense of direct communication. Although all animals possess and use this sense, human beings, who were produced with a very developed brain and psychic system, do not use this sense much anymore, and instead tend to close in on themselves in order to develop their personal thoughts.

Here, we present the problem of this deep organizational comprehension between living beings, which engage in the network of the informational substrate to use it.

7.3. The sense of informational comprehension or the sixth sense

On Earth, every living element has an informational envelope that links it to other elements. The links are natural, generated by the creation and activity of the elements, thus realizing the union at the informational level of all that is living. Every link is an informational link that follows the unifying characteristic of the organizational law, using the substrate of the space components. The information that propagates is indicative of the organization of each living being, and can be used by the solicitation of an organism that wants information about an element of its own world. The sixth sense, which we call the sense of informational comprehension, will be the informational action of taking information enabled by the informational links between informational envelopes of organisms, with the links being activated between the informational envelopes in order to realize the solicited communication.

We will also assume that this informational comprehension is in fact common, that it is performed almost constantly in animals, and that its voluntary use to comprehend the state of a distant organism is a specific case produced with explicit intent. There will be two cases of use of the sixth sense: an implicit use that gives contextual information and a voluntary use, where we must focus on a subject to receive information from its informational envelope.

7.3.1. *The implicit communication process of the sixth sense*

It is common to simply comprehend things through the sixth sense. Animals and humans immediately understand the atmosphere of a group in their environment. There is information about the state of the group, about its dominant or abnormal nature. There is an comprehension of the friendly, pleasant or problematic nature of the group through the dominant morphological pattern of the group. After this informational comprehension, emotion is then triggered, followed by the cognitive process according to this emotion. The process of life in its environment is thus normally grasped through informational comprehension, followed by sensitive, then cognitive comprehension. Animals systematically practice this, with reduced cognitive comprehension. They live in environments where they comprehend the dominant organizational information, allowing them to limit taking too many risks in the presence of predatory groups and ultimately survive. They form groups according to the informational relationships between members, recognizing a dominant member as necessary, ensuring the organization and maintenance of the group.

However, this way of life is not the case for humans in the Western society. During childhood, humans have systematically learned to use their cognitive representations in order to communicate through language and to structure their activities by habit. In their life, they tend to deploy their psychic manifestations in these cognitive forms that use memory and especially language, which has a very deep organization. They can therefore live and work in elaborate mental reflections, especially with smartphones, which tend to make them close in on themselves. They can thus consider that they do not use the sixth sense and even quite often assume that it does not exist.

This is not the case in the traditional societies of Africa and Asia, where the comprehension of the environment, living things and human groups is a fundamental cultural characteristic. Western society is based on a materialistic approach to events and to all reality, which is considered to be absolutely measurable and decomposable, and therefore knowable and predictable by measurements, and the notion of immediate informational comprehension is hardly considered. However, this sensitive comprehension of the environment by informational fields can be further developed by a voluntary use of informational fields.

7.3.2. The voluntary process of informational communication

We have well established that an organizational immersion regulator in an informational state exists in the psychic system, which can be used to enter in direct informational communication with the external informational envelopes of other entities. It will thus be necessary to activate this organizational regulator to comprehend a specific informational envelope.

Analysis of the usage process of informational links by living organisms is carried out according to two actions: the establishment of an available informational link, followed by its use. We will call the individual who requests a specific piece of information through the informational substrate the *requestor*, and the other individual who is targeted, the *receiver*, who receives the informational request and who will automatically send the requested information that is present on their informational envelope. The process is as follows:

– There is the establishment of an informational link by selection in the set of links, which will become permanent between an individual and another well-known individual, or between the requestor and a specific geographical domain, which will allow the individual requesting information on the state of the receiver to be in connection with them, and then to receive the general information given by this receiver.

– The informational link, which is available through a request for an encounter and the assessment of the state of another well-known individual, will be activated through the informational envelope of the requestor, taking the basic information about the organism that is thus linked to them on the informational level. There can be absolutely no dialogue at this level, only transmission of the organizational states of the organisms, which communicate by action of the informational link.

– We can consider two very different cases of using this information link:

a) There is the capacity of activation of the informational link that has been generated by the requestor, allowing the receiver's transmission of organizational information to know its global state or possibly to locate it geographically.

b) There might be the requestor's capacity to use their informational envelope as an elementary control center, so as to send an informational form of the dominant morphological pattern on the informational envelope of the receiver, influencing it. This is another problem altogether, because some of the aspects refer to witchcraft, which is not scientific at all.

The problem is the use of the informational link that exists between all organizations. The questions are then the following:

1) Do living organisms that meet, live together and appreciate each other, all have active informational envelopes that activate communication?

2) How does the informational establishment of a specific link that is identified by one organism with another organism take place?

3) How does the selection for the use of this link that has been identified by its psychic system take place?

4) If the link exists and is active, how does the comprehension of the received information take place in the psychic system of the requesting organism and what is its level of comprehension?

The answers to these four questions will give the solution to the use of the sixth sense. The process that will momentarily link two organisms together through a strictly informational relationship, between a requestor of information and a receiver who will transmit a response, is carried out following the five steps that we will describe.

7.3.2.1. *The five stages of voluntary comprehension by the sixth sense*

7.3.2.1.1. Step 1: active existence of the informational envelope of living organisms

There is an informational envelope for each living organism, which is a synthesis between its external corporeality and its informational layer, as well as an informational envelope in the brain communicating with the envelopes of the body's senses. This envelope is the domain that establishes the informational links between the organism and the whole living system, which is external and familiar to it. If the organism is in good condition and is not in a pathological state, there is normal functioning of its informational envelope, which is linked to the informational substrate of the world. There is also an envelope in every structured and organized set of the world, in the places where living organisms live socially.

7.3.2.1.2. Step 2: establishing an informational link between two organisms

The creation of an informational link between two individuals, a requestor and a receiver, is done in the following way:

1) First of all, there is a physical meeting of the two individuals who comprehend each other using their five usual senses, who appreciate one another.

2) There is a comprehension and willpower from one or both of the individuals to communicate with the other, so as to be able to follow them, to comprehend them, to know their state and their spatial position. There is therefore voluntary activation of the informational regulators of the two organisms, allowing the communicating action of their informational envelopes.

3) There is then a permanent existence of this direct informational link, wherever the two individuals are located geographically.

4) This is all similar when a requestor lives in a specific area that they like, and wishes to have information about this area, when they want it.

7.3.2.1.3. Step 3: the request and activation of the informational link on the receiver

In the informational relation, there is the questioning of the psychic system of a soliciting organism, which will focus its attention on another organism that it has good knowledge about: the requestor wishes to comprehend another organism at the informational level, without the use of the five usual senses, which do not operate in this process. This is a specific aim. They realize the request, which appears as psychic willpower focused on a subject, before their informational envelope is put into action in order to realize the putting into action and the use of the informational link that connects them to the receiving organism, which they wish to have information on.

The link between the requestor of information and the receiving individual is established by the decision to voluntarily put the informational regulator of the requestor into action, so as to activate the informational envelope of their brain and to comprehend the informational envelope of the receiver through the links carrying information.

There is a conscious request from the requestor about the state of the receiver. The informational link is selected by the specific aim to become an action and by the requestor's thoughts, who directly transmits this request to their informational envelope. This is a major process that is not very common, which we will develop.

There is the transmission of a request of the general state of the receiver by the requestor's envelope, which is a simple query-type request concerning their global state or their spatial position, and it is never a specific, precise question. This request will be conveyed by the informational link.

7.3.2.1.4. Step 4: the information transmitted by the informational link

The receiver receives the query on its information envelope and reacts by transmitting its global state in response, through the general regulator of its information envelope.

The information generated by the informational envelope of the receiving organism is a strictly informational and morphological form, opening onto the main characteristics of the organizational state of the receiving organism.

These main characteristics of the receiving organism are informational and quantitative; they specify whether the receiving element is in a good or bad situation, whether its current action is good or in a weak position, etc.

The physical distance between the requestor and the receiver, as well as the geographical direction where the receiver is located, will be globally specified as a directional indication and a distance amplitude in the transmitted information, giving the characteristics of the organizational relation of possible coactivity, or not, between the requestor and the receiver, which is an operational rule of the organizational law.

7.3.2.1.5. Step 5: comprehension of the transmitted information

The informational link has thus allowed the requestor to receive basic information about the receiver's organizational state.

This information is simple, practically elementary; they are small sets of informational fields. They will be transmitted and expressed in the neuronal system of the requestor by their informational envelope, thus allowing them to comprehend them in a sensitive and conceptual way. There is therefore an interpretation of the basic information in the form of generation of indicative neuronal aggregates in the neuronal system, making it possible to generate the basic elements of psychic comprehension, which will be tested.

7.3.3. *The solicitation of the sixth sense and the comprehension of the informational substrate*

We must now clarify how the action of the informational regulator, which activates the informational envelope of the brain, is solicited.

7.3.3.1. *The cerebral domain of the sixth sense of perceiving the informational envelope*

The brain of living organisms possesses a particular domain that is made up of specialized neurons, allowing it to comprehend its informational envelope, to activate the informational regulator, to locate an informational link in order to memorize it and make this memorized informational link active, to receive the information transmitted by the link and to transmit this information to the specialized neuronal system, then to the central neuronal system, which conceptualizes it in order to generate a mental representation. This domain of comprehension of the informational envelope, which we call the cerebral domain of the sixth sense, is in the periphery of the neuronal system, in its physical envelope linked to its informational envelope. It can only be made accessible and be in noticeable activity if the neuronal domains representing the five senses are no longer in action, and the neuronal system focuses on its use by its informational regulator. This domain has the form of a neuronal graph, placed at the periphery of the brain. Elements of this graph represent the memorization of links with certain organisms, and other elements can comprehend the information received and generate a corresponding neuronal form for the use of the conscious mind.

The existence of this sixth sense cerebral domain can be seen as a normal occurrence by the exercise of the organizational law that created living things on Earth, by massively using the informational substrate to organize everything in a communicative way. It created autonomous organisms with numerous internal informational envelopes and had to provide these autonomous organisms with a means of communicating at the informational level, through the use of informational links. It therefore generated a domain in brains that allowed any organism to communicate, so as to comprehend the organizational states of other organisms or the state of specific geographical and physical domains. It is the development of the living organisms that would lead to the evolution of the capacities of these domains of informational communication, by allowing the evolution of mental comprehensions. We will see that the case of human beings is a little contradictory, as they practically do not use their sixth sense cerebral domain anymore, and instead mostly internally use the considerable conceptual capacities of their brain.

But how does the informational regulator that initiates the action of the sixth sense come into play?

An informational message is a reduced dynamic form that exists in the informational substratum of the Universe. It is a morphological characteristic constituted of basic elements of the substrate, which describes the quality of the organizational state of an element on its informational envelope. To receive this morphological indication, it is therefore necessary for the organism to be situated at the level of its informational envelope, to be at the level of its basic informational elements coordinated by this envelope, and to no longer have sensitive and cognitive references to space and time. The organism must first focus psychologically on the identity of the receiver it wishes to comprehend, establishing them as the subject of the informational request and, for a short time, voluntarily annihilating all the subjective impressions it may receive from its psychic system. This amounts to them being totally focused, as a conscious organism with a body endowed with an informational envelope, on the informational subject they want to comprehend. This is a posture that is natural for all living organisms, it is a placement of their psychic system of comprehension at the informational level, using their informational envelope and their sixth sense cerebral domain.

7.3.3.2. *Informational comprehension by the sixth sense*

To place oneself in ideal generation at the level of the informational envelope of the brain involves not using the five usual senses that participate in any mental representation with the underlying or explicit comprehension of the spatial position, of the time and of the designation of things. It is a strong decision to position oneself in the informational space, with a radical cut-off of the comprehension of the usual space-time. The mental system must control and put its informational regulator into action, allowing it to activate its informational envelope with a choice of links. This mental potential is present in all animals. The very specific mental system of human beings, which is usually closed in incessant sensory and cognitive developments using language, tends to inhibit this option, which disappears in childhood education, and they have to do a great deal of mental work to recover it.

This process of choosing an informational comprehension is done by the functioning of the following mental system:

– The will to fix the subject of the next mental form on a specific object of the world, with the cutting off of the comprehension of the usual senses and of the mental conception, for the informational comprehension via the cerebral domain of the sixth sense, which is then activated.

– Activation of the informational link in the cerebral domain of the sixth sense according to the aim, launch of the direct solicitation, which is a simple query, then reception of the informational form that is automatically transmitted by the receiver.

– Cut-off with the informational system and analysis of the received informational data, so as to make the corresponding mental representation comprehensive sensibly and conceptually.

7.3.3.3. *The informational substrate produces physical space and an informational geographic map*

Perceiving the informational substrate through the sixth sense develops a mental landscape consisting of the informational elements of the substrate, which shows the following:

– Their close existence during the displacement of the requestor. There is a geographic location of the distance and of the possible proximity of the targeted object, through the assessment of the link between the perception of the sixth sense and the intensity of the symbol point.

– Their state during informational contact, to locate and feel the organizational state of the targeted object. There is an comprehension of the energetic state of the targeted object through the comprehension of its informational envelope.

The informational substrate is the space-time of our Universe. To go from one point to another means to move both in the physical space and in the informational substrate. Therefore, to be at a certain place in the movement between two points of space also means to be on a sequence of informational links between two informational envelopes, which give information about the organizational states of the two points, about their situations.

Moving towards a targeted receptor allows us to comprehend the sequence of informational links that lead to it and to follow them. Thus, for a dog who is looking for his master who has just been hospitalized, this would lead him to follow the informational links that go towards the latter's informational envelope, without breaking contact with him, knowing how to move towards this envelope by taking unknown streets in the city and ultimately reaching the hospital park, and feeling his very close presence.

7.3.3.4. *The informational substrate generates physical space and an informational geographic map*

The informational substrate is physically directive, giving a general comprehension of the known sites, giving a sense of orientation, direction and the distance between a requestor and a receiver. It provides the perception of a set of distant and possibly remote informational sites, constituting an informational geographic map.

It is clear that migrating swallows have a very sharp perception of a target site that is considerably distant, but which is a dot among others on the mental map of informational sites that lead to it.

The perception of informational links takes into account the physical distance between the requestor and the receiver, by giving the energy relevance of the targeted informational envelope. Being very close to the receiver is a potentially intense contact of the informational envelopes, while being far away shows weak information between the two elements.

The problem then lies in defining how the informational form received on the informational envelope is interpreted by the neuronal system, which obviously depends on the informational characteristics composing the form.

7.3.4. *The cognitive and sensitive interpretation of the information forms received*

Once the relation between the sender and the receiver is established by the informational link, there is a systematic transmission of the organizational state of the receiver, provided by its informational envelope. This is the seizure and the transmission of the very general characteristics of the general pattern of the informational envelope of the receiver, which arrives in the informational envelope of the sender.

This information, which comes from the general pattern of the receiver's envelope, is a small set of informational components, with some specific patterns that make up the structure of the available information. This information will be transmitted in this form by the informational link, then received in the informational envelope of the requestor, before being directly transmitted to their sixth sense cerebral domain. It is thus a direct communication between two informational envelopes. Furthermore, the interpretation of this information by the requestor leads to strong activity of the neuronal aggregates by the neuronal system of the sixth sense domain, corresponding strictly to the informational drives received in its informational envelope. There is a neuronal interpretation of the transmitted message by the link on the informational envelope of the requestor. The neuronal system knows how to interpret all the energetic, visual and auditory information, and it also knows how to interpret the forms made up of informational components that have been received in the cerebral domain of the sixth sense.

The cerebral domain of the sixth sense has the following form:

– a dynamic graph structure on the periphery of the neuronal system, whose nodes and links change;

– there are, in fact, two kinds of nodes in this graph:

- informational components that are active upon reception of an informational message and that form a small sequence of informational fields to capture the transmitted informational components. They are interpretable as sets of nodes of the graph,

- the nodes establishing links with neuronal aggregates of the brain's central system, so as to transform the informational message received, which is in the form of informational fields, into a mental form made up of aggregates of active neurons, performing cognitive and sensitive interpretation.

The complete activity of this cerebral domain of the sixth sense is therefore the following:

1) Activation of the cerebral domain by solicitation of neuronal aggregates.

2) Action of the neuronal aggregates of the domain soliciting its activity by the voluntary request.

3) Waiting for the reception of the message coming from the receiver by the activated informational link.

4) Acquisition and interpretation of the received message, which is a small sequence of informational fields, by the activity of a set of informational components of the sixth sense cerebral domain.

5) Formation of a set of aggregates representing the neuronal interpretation of the form.

6) Action of the set of patterns and neuronal aggregates of the usual system of conception of the representations, so as to interpret this new neuronal form in a cognitive and sensitive way.

7) Formation of a set of neuronal aggregates in the conscious mind, representing the complete interpretation of the received and totally decoded message, in order to generate new analysis representations of this comprehended form.

There is a reading and analysis of the informational message received on the nodes of the graph representing the domain of the sixth sense, each active node then being interpreted by neuronal aggregates that are in constant connection with the sixth sense domain. This is a classical interpretation of external information at the cerebral level, which is a sequence of informational fields of the substrate of the Universe, as it is also what the five usual senses do, which analyze messages received in the form of photonic or phonetic fields, as are the interpretations of images or sounds.

We can therefore say that the sixth sense is the comprehension of the informational substrate.

7.3.4.1. *The comprehension of the informational substrate by the sixth sense*

The creation of the Earth is based on its informational substrate and its continuous informational field, which conforms and organizes all its physical elements. The sixth sense will be the latter's comprehension and use. It will allow the one who uses it to comprehend the organizational and spatial situation of multiple elements, by using the informational links between them and these elements. They will be able to locate the spatial paths to reach them and receive information on their organizational states. This will allow human beings, if they can use this sixth sense, to understand the complexity of the organization of the elements of the world and their links, to comprehend the complexity and the beauty of living things, and especially to directly communicate with many other humans and animals, reinforcing their core values.

It would be extremely necessary for humans to use their sixth sense in order to understand their shared and constructive place throughout the world.

Furthermore, comprehension by the sixth sense produces an informational mental representation.

7.3.4.2. *The informational mental representation*

Animals that can comprehend through the sixth sense construct a specific representation in their mental system, formed by the action of the patterns of the sixth sense cerebral domain, which produces a neuronal layer where the informational state of the targeted person, or the physical site and its distance, appears in the case of access to a well-known place.

This mental representation is very different from the representation given by the comprehension of reality that is perceived by sight, where the landscape is comprehended as composed of multiple elements. However, in both representations, there is a distance assessment relation which is, in the informational representation, the representation of the actual physical distance between the perceiver and the thing comprehended. This is how dogs in a car know that they are going to get home quickly, without seeing the surrounding landscape. This notion of distance in the informational representation comes from the informational substrate that makes the physical space, and whose links between the informational envelopes provide a distance value that can be comprehended. For the living being that can comprehend in an informational way, this can be achieved by following the characteristics given by the informational envelope of the Earth, making it possible to represent the real space-time and its local metric.

7.4. Common use of the sixth sense

We are going to describe the very frequent use of the sixth sense in animals, as well as show why human beings do not use this sense, by mainly developing their production of complex internal reasonings, building on their considerable rational and management developments of comprehension, and the analysis of their areas of life.

7.4.1. *Common use of the sixth sense in animals and humans*

There are living beings who have precise informational comprehensions about the states and situations of other living beings, as well as about the states of physical things that interest them. The sixth sense is to comprehend things in a certain way that are not directly felt by the five usual senses. It is a particular comprehension of objects of the world that are very often out of the reach of the five usual senses of the requestor of information. We have put forward that this comprehension is based on the use of the informational substrate, which links all the informational envelopes of physical elements and living organisms, linking the envelope of the receiver and the envelopes of the elements that they know and can comprehend. We assume that this comprehension is a specific mental form that is generated in the psychic system, which is then mentally assessed, thus leading to emotional and conceptual generations.

The links between the informational envelopes of the elements are direct informational lines. Each informational envelope can be connected to another by physically meeting through a direct straight line, thus linking the two organisms which can move, with the informational lines being under space-time and which can be continuous. They represent the cooperative or negative influences between organisms, often being the links of cooperative and evolutionary organizational activity of groups of living organisms. Therefore, only the living organisms that have had a community of action, that have met and been in the same community for a certain time, will be able to communicate using these links. Furthermore, once the informational link is established between two organisms, it remains present, regardless of the positions of the organisms that may move.

We must then assume that living organisms have permanent informational links between them, because they live together, and that the comprehension of the envelopes and their meanings through these links can be very often comprehended. For comprehension to take place, the organisms must be in direct communication for a moment.

It must be clearly stated that animals have this specific operational sense, unifying their position and state with the informational substrate of the elements of the world, giving forms of information about specific things that are in a large spatial spread. They comprehend living beings, whose informational envelopes provide information about their situations and movements, and also about the state of well-localized areas by informational envelopes located in an area that is perceived as important, because it is known: to be far away and to move away, or to move closer to an element that is considered as important. This comprehension can then be conceived as a situation of mobility, which is appreciated as such in real space-time.

Many animals that live in the wild can sense that bad weather is coming. It is not in any way a vision of the future, but the comprehension of bad situations experienced by other animals in environments that are close by, who have experienced and comprehended bad weather that has occurred in their domain by describing it on their informational envelope. It is therefore an comprehension of meteorological phenomena through the perception of what is happening in the vicinity, on a recurrent basis. This can therefore happen quite far away, through direct communication with other animals.

We must assume that an animal initially lives with its continuous informational links with other animals, so as to form groups. Wild animals form groups with a dominant animal that perceives the behavior and state of the other animals in the group through its sixth sense, thus managing the organization by intervening with an animal that has a behavioral problem, when necessary, which will have been indicated through information on its informational envelope.

We can use the example of migratory birds that possess the ability to fly over a thousand kilometers from one nesting center to another. In order for these birds to find their way every year without problems, they have to follow an informational line between the two sites, which they perceive without problems and which, for them, is the informational route of their migration. Bird breeders have described games with pigeons that are taken at night in trucks, far away from their usual nests, which are released and then fly back to their nests. They all find their nests, but the one who arrives first is considered the winner of the game. It is absolutely impossible that these birds use the Earth's magnetic field to go to a specific place, since the magnetic field does not give any direction on the Earth's surface but that of magnetic domains. We must therefore assume that there is another field, and we clearly assume that it is the informational field.

This connection between the informational envelopes of animals is always realized, because the substrate of the Universe is a dynamic hypergraph of informational links that generates continuous sequences of large links, permanently linking organisms when this link is requested by one organism and accepted by the other. We must therefore admit that animals have the capacity to generate links between informational envelopes that become permanent, regardless of the physical positions of the animals in space, when the element considered by the requestor of the link enters their memory, which is connected to their sixth sense cerebral domain.

It is possible that a group of animals using the sixth sense on an isolated animal with negative intentions will actually result in negative changes in the animal's condition and behavior. This can happen in groups of animals that are still used to regulating the behaviors of group members, and do not like animals with irrational behavior that destabilize the group. This would be a global influence of the members of a group against an isolated case, an operation similar to the creation of a dominant morphological pattern sent by the substrate, in order to reduce the behavior of a negative individual in a well-observed group.

There was indeed an evolution of mammals that produced human beings, a continuous constructer in the world with their particular psychic system, since they constantly generate mental representations that they analyze and reconstruct by rethinking their experienced thoughts. They live and constantly build physical things in their usual practice, so as to generate their society and, therefore, their civilizations with intense and continuous technological developments.

Human beings are, most of the time, in practice where they conceive themselves, by carrying out activities that they know how to psychically envisage. They generate, experience and assess comprehensions that come from their five senses and they produce internal reasonings almost always using linguistic elements, which are very complex and prominent forms in mental representations. They continuously assess the sensory and conceptual comprehensions they develop. They always know that it is their "I" that make their thought. This posture, based on continuous internal reasoning, leads to a kind of confinement on the "Self", a concept that is considered as significant and central in practically all cultures, as well as in most philosophies. This self-confinement leads to a certain disruption with the informational link, which is a direct comprehension of everything through the use of the cerebral domain of the sixth sense.

However, there is a case where humans uses their sixth sense. In order to comprehend through this sixth sense, they must put themselves in a position where they temporarily cease to use their sensitive comprehension, mental reasoning and linguistic memory, opening themselves to the informational world without using their memory of present and past events. This comprehension occurs when they think of a person they know very well, without referring to memorized events, simply by thinking of this person in a strong way. They feel their presence, which may be distant, they comprehend their informational envelope through their sixth sense, knowing that they are in a certain place, somewhere and that they are alive. This sensation completely ceases when this person, whom they comprehend directly, has just passed away. In the latter case, they feel a real void in the comprehension of this person; there is no longer any comprehension of the informational envelope that characterizes the living being that they frequently comprehended. There is a disruption and their sixth sense can no longer comprehend the envelope of the person who has died and disappeared. They feel their disappearance from the world through a comprehension of emptiness. In order to think about them again, they must then call upon their memory to represent memories. There is therefore a continuous informational link between a person and the people that they know very well and are used to seeing. This proves that the informational network is indeed operating between the informational envelopes of human beings, just as they are between the informational envelopes of animals.

Furthermore, there is a development of informational links in people who are together in specific places, like churches or synagogues. In these places, which place humans as very small elements in a closed and immense volume, they are in a state of domination in the world, yet they open up to direct communication with others because they follow a rite, without analyzing it. This direct relation between participants has made the development through the action of religions.

However, today human beings often live closed in on themselves, where they deeply conceive their social and cultural practices to be realized. They have founded their very structured civilizations based on this mode of being, with closed regulating domains that realize the order of things. They do not often think of direct informational comprehension with others, so as to comprehend them as active and necessary for socialization.

7.4.1.1. *human beings closing in on themselves*

Human beings have brains with considerable organizational capacities compared to the brains of other animals, which has allowed them to develop languages with scripts. They can conceive a thing and question it using their memory, and then question their answers and their own questioning by analyzing them.

This means that they have a very great capacity to generate thoughts internally, by placing themselves in their ideal world, which has a considerable linguistic substrate. This position is a posture of internal closure, which will be the cause of the disruption of action of their sixth sense cerebral domain with the universal informational substrate. The comprehension of the informational links of the substrate is direct; it is an opening without conceptual questioning for this comprehension, only having to be active for a short time. Society has made human beings frequently live closed in on in the Universe of their sensations and of their conceptions with sentences made up of their languages, internally questioning all things.

We must assume that human beings do not make their cerebral domain of the sixth sense available in their brain. Their brain usually remain closed on itself, generating conceptions each time they feel and think, using highly organized memory and language forms, which is a considerable set of conformations of their dynamic memory. The sensations produced in humans are almost always conceptualized using words and sentences, these conceptualizations strongly following the life and death drives defined by Sigmund Freud (1966).

Today's human being has designed a general computer network that covers the planet, allowing everyone to communicate through words and images with a number of other people or software systems, which is a technical achievement that has a certain relationship with the informational substrate.

However, although the use of the sixth sense is ethical and engages those who practice it to a distributed fraternity, the computer network of our unequal world, where economic confrontations do not cease and where war is practically perpetual, is not going in this direction.

7.4.2. The action of the sixth sense for hypnosis, the power of magnetism and meditation

We will now see that the sixth sense can be used to exercise control over other people.

There is a way to control another living being: by putting it in a state of artificial sleep with domination, which is what hypnosis does. Hypnosis puts an individual in a state of inaction, of artificial sleep, for a certain time, so as to be able to proceed to perform actions on them.

This state can obviously be induced by specific drugs, but it can also be induced by the hypnotist's own will. In the latter case, the two people are in close proximity, observing each other, where one is in a position to control the other with their acceptance. In reality, the hypnotist links their informational envelope with that of the individual who is in close proximity, and whom they identify very well. They establish a dominant informational link with this person, with the firm intention of making them pass into a state of inaction. It is thus the establishment of the strong control of the informational envelope of a person on another, which is achieved by agreement of the dominated person who expects to enter the phase of domination. The person's unwillingness to be hypnotized is done by exercising their will to stay awake, as well as their confinement in their psychic system, thus refusing the opening of their informational envelope. Once the dominant informational link has been established and the hypnotist's control of the informational envelope has been established, the hypnotist can proceed with physical actions on the body of the hypnotized person who is no longer reacting, who is in a state of artificial sleep.

This practice of taking temporary dominion may represent the activity of traditional healers, who use what is called the power of magnetism. The term *magnetic* comes from the existence of earth magnetism, but it is not at all this magnetic flow that healers use. There are cultures that state that humans have the ability to communicate directly with others near and far, as well as to send them powerful information that changes their behavior. Humans who are able to do this are, in traditional cultures, called healers and sometimes witch doctors.

The notion of magnetism is considered, at this level, as a flow of force that allows energy to be diffused on another person, in order to modify their state. It is a notion that has always been used, from time immemorial, in human cultures. Using this property of magnetism, it is considered that healers, also known as magnetizers, use their power of energy transmission to relieve other people's ailments, by passing their fluid through their hands, used at a small distance on the sensitive parts of the body of the patient. Some consider this magnetism to be a flow of electrical vibrations and radiation emitted by the healers' hands. However, the basis of these activities is totally different, because every organism, every person, has an informational envelope that they can activate in themselves, allowing them to communicate at the level of their informational envelope with other people, and thus establishing major sets of informational links that send out morphological patterns. This ability to communicate through flows of pattern-sending informational links explains the ability of healers. It is therefore not magnetic flow from the healer's own energy that relieves the pain of a patient in close proximity, but a specific use of the informational flow from their informational envelopes. Healers would thus have the ability to use their informational envelopes to transmit morphological patterns into the informational envelopes surrounding the organs of patients, relieving their pain. This is a very strong ability.

This ability to remove pain in people is a deeply social and humanistic ability. It is based on openness towards others and the desire to do them good. This should certainly be generalized in our type of societies where individualism is developing very strongly. As very well specified by Sigmund Freud, there is a life drive in human beings that leads them to communicate with one another and to appreciate each other, in order to provide and receive help. He also defines another drive he calls the death drive, which leads to the desire for domination over all others and confining oneself for one's own pleasure. This very dark drive has led humans to engage in killing activities throughout all their civilizations and it has led some, who are deeply lacking in humanistic culture, to assume that one can reduce and even destroy another by sending negative forces from a distance, forces that come from "beyond the world".

It may be possible for large groups of humans who simultaneously think negatively about another person to activate a negative morphological pattern about that person. If this were possible, it would be good to control such human groups so that they do not all organize themselves to simultaneously think negatively about altering another person.

We will now discuss the question of meditation. We can consider that the practice of meditation is, in any case, a contact with the informational layer of the world, which involves understanding that we are a very small thing in the organized immensity of the world and the Universe, but that we can open ourselves to the problem of the extent of the world. By practicing meditation, or by engaging in a situation of sharing with living things, and in particular with plants, humans can move towards ethics by making their life drive hegemonic, through learning to use their sixth sense.

For the approach to meditation that has been described as transcendental, we can refer to the work of Maharishi (1976). According to his work, transcendental meditation must allow the individual to open up to the world, in order to share with everyone. He develops a conception of the Being that he presents as an absolute that the mind can reach, by leaving the concrete and local world of his everyday life. This can be seen, in our approach, as a communication with the informational substrate that immerses the world, organizes it and makes its elements. Maharishi posits that the all-overseeing Being is an eternal absolute and can become Spirit or Objective Matter as the case may be, and thus form and transform the world, which is the ability of his divine character. There is this Supreme Being and there is his instrument, which is karma. Karma is the instrument of the Being and realizes reality through creation, evolution and dissolution. In our model, we have presented an organized construction of the Universe, with the informational substrate that is the informational space underlying the structured reality, which is produced from a generative component. This substrate constantly acts to initiate the creation of all material organizations according to the organizational law, so as to continue the development and organization of the Universe, and of our planet. We cannot pose the problem of an eternal absolute, as it cannot be comprehended in our model.

In Maharishi's explanation, all thought is constructed at what he calls the deepest level of consciousness and appears on the surface of consciousness to be effectively felt. In fact, his approach leads to the assumption that thought is generated at the level of energetic and informational neuronal aggregates that are organized, some of these aggregates, through their envelopes, being in communication with the informational substrate. The fact of producing thoughts in the mind is a particularity of the living being that has been created through its evolutionary organization, and we can assume that, through its informational elements, this substrate always accesses and participates a little in the creation of each thought. The problem for human beings then lies in accessing this informational substrate.

What Maharishi advises in his transcendental meditation training is to access the minimal level of thought, in other words, when it is under construction with informational communications with the substrate, to enter into communication with this substrate which is the generator of the Universe. By accessing this level of generation of forms of thought, which he calls the source of thoughts, we are at the level of communication with the substrate of the Universe, we become aware of this substrate's extraordinary power of construction to generate an organized Universe. By understanding his posture of comprehension before the generation of the elements of the Universe, this should give the person who accesses it a path to tranquility, happiness and peace, through the informational substrate guided by the organizational law.

We posit that the informational substrate, which makes physical reality, can act at the level of thoughts in realization; it can influence thoughts that are organizational movements of neuron fields, morphological patterns and informational regulators in brains. People's thoughts can communicate with one another at the level of elementary informational communications by the sixth sense, but are not influenced in their conformations producing ideal forms by virtual forces, coming from a hypothetical absolute designer of the Universe. Our model assumes the existence of a continuous, systematically expansive generating diffusion that makes space and all the elements be at all scales, all subject to the organizational law. There is a well-defined beginning of this process, a general rule of development and constituted states whose creation, behavior and general transformations are known.

We must consider that the creation of the multiple living beings, all the way up to human beings who have the capacity to think thoughts, are considerable developments. This requires an absolute respect for life and the development of social humanism in societies.

7.4.3. *The notion of premonition and the sixth sense*

We will see that we can use the sixth sense to comprehend trends in the very near future.

There are researchers who have done a lot of work on premonition, showing through experiments that humans can perceive an event before it happens (Damasio and Bechara 2005). This has been studied in different experiments, such as risk situations, where humans choose a solution by removing the dangerous case, and card game situations, where humans have to choose goods cards from a face-down pile, sensing negative cards for their games and opting to not take them.

Researchers deduced that humans could foresee the very near future in certain situations and that the notion of temporal movement was then to be reviewed, as well as the fact that humans had a notion of premonition and that they could access images of future situations that could unfold. Some have even posited that time in the world is not linear, systematically going from the present to the future that does not yet exist, but that the future can be somewhat comprehended (Hardy 2012). This will not be our position, and we will still refer to informational links to present these situations.

Human beings are always in a situation in the present world, where many morphological patterns are active in their environment and are accessible to them through informational links. These patterns concern all physical objects and all living beings through their informational membranes. We have posited the existence of morphological patterns of assessment defining the most probable future situation for the substrate, in a domain well delineated by the action engaged in this domain. This laid down the existence of a possibility of knowing the very near future of any situation, which is engaged by the informational conformation given by these morphological patterns.

For a human being in a specific situation, the comprehension of an evolution and a very close future is therefore achieved under the following conditions that we will describe.

7.4.3.1. *Comprehension of the very near future*

In a group of humans or animals that are in strong communication for the management of a problem considered to be important, individuals may experience strong questioning with strong emotions about the evolution of the problem. In their informational envelopes, this can generate the activation of morphological patterns on this sensitive and strong questioning. These patterns can then communicate with the patterns of regulation of the substrate which, in this case, generates an assessment of the near future, in order to be able to regulate it and to organize the stability of the behavior of all the individuals. This communication between the patterns of questioning and regulation will then, through the sixth sense of the one who is questioning, be able to give an comprehension of the near future and of the actions that are unfavorable or favorable to them. It is in this way that they will have premonitions, allowing them to manage the positions to be taken and the actions to be carried out in the situation.

We have thus assumed that we are in the functioning of a group where the situation is estimated by at least one participant as being very important, through dominating their emotions and affectivity. This engages the informational components to comprehend this situation as having a lot of magnitude for at least one individual, therefore specifying the generation of assessment patterns. The response to a person's intense sensitive questioning thus engages the informational substrate to generate particular assessment patterns. If the person is highly polarized on their questioning, does not reason, does not use language and its formalisms, and just uses their sixth sense alone, then they can directly comprehend the assessment patterns in their informational envelope and transform them into a clear mental assessment. This is very explicitly the case of an comprehension of the most probable near future.

The comprehension of a thing is therefore a particular state of informational communication, where the one who comprehends can psychologically engage in a position that allows them to use their sixth sense. Let us note that there is no possibility of engaging in the informational system with a particular psychological state in order to harm another person, which is what sorcerers claim to do. Access to the informational system underlying the real world only means possibly accessing some of its information given by patterns, which can then be cognitively interpreted.

7.4.4. Thoughts and the safeguarding of the world

Our planet has entered a process of species extinction. There are 85,000 species of animals studied, and it is believed that 25,000 species are extinct. 26% of mammals, 13% of birds and 43% of amphibians are currently in the process of extinction. The cause of this collapse of life is the suppression of plant spaces, as well as the considerable and continuous pollution carried out on Earth and in the seas by the industrial society. Human beings no longer live in harmony with the plant and animal world. In this very serious climatic and ecological collapse in which we are engaged, we could say that "wildlife is the only thing capable of creating diversity to adapt to disruption" (Sarano 2017). Nature is disappearing and being replaced by huge areas of housing, commerce, transport, industry and technological production sites, where products are manufactured, many of which are toxic and generate innumerable piles of waste. It has been calculated that every second, 100 tons of waste are dumped into the oceans, and most of it is plastic.

Today human beings have become individualists, locked up in their dark impulses committing them to their own pleasure; they are well subjected to their death drive, as defined by Freud (Freud 1966). In general, they no longer know how to live openly, peacefully and quietly and, above all, how to share communication with others, with natural plants and animal nature. They no longer know how to live by acting quietly and having great moments of meditation on the beauty of life. They live with ceaseless actions that they constantly get faster at, in technological spaces that are developed without ethics. They have overpopulated the planet, with more than 400,000 births per day; they have transformed natural spaces into urban and industrial sites, where the wild animals that were on Earth before the human species lived in harmony, in the magnificent natural plant world. Human beings dominate absolutely all the planetary space, but they do not dominate their reflections nor their actions to live their life peacefully, thinking that they are sharing with all that exists.

All animals use their sixth sense. They know how to finely organize themselves in their groups by using informational communications, by finding animals in difficulty, so as to help them. Furthermore, the law of organization always intervenes in reproduction, adapting groups of animals to their environments, eventually making changes when the environment becomes deficient, which is a regulatory process that requires time. However, it is with this law that wild living species will be able to adapt to current considerable changes in the ecosystem during their reproduction, although some adapt much better than others with an organism that allows it.

In this world, it is very important for humans to be able to communicate by using the sixth sense, as well as to constantly adapt to each other by forming well-organized groups and to share moments of difficulty in life together. This would be a considerable transformation of our type of life in our societies which practice confrontations, battles, conflicts and wars, because there would be elementary communicative sharing, moving towards fellowship and the systematic helping of all those who need it. We would directly feel who is in difficulty, and how we could help them to improve their thoughts and situation. Therefore, can we believe that applied, sociological or even educational research can be undertaken to make the sixth sense effective in all humans, so that they practice it? Our society has transformed fundamental research into multiple technological developments, whose central rule is competition with the hegemony of success. Here, it would be absolutely not a question of competition, or of dominating other groups, but of establishing global sharing and universal ethics in a peaceful world. This would be a very profound transformation of society.

Conclusion

It was necessary to develop a unifying model of the creation and development of the Universe and living things in order to deal with the sixth sense, which is strictly informational. The only unifying model is based on the notion of informational flow that is realized by the organized expansions of informational components, forming the substrate of the physical Universe. To assume that the Universe does not come from anything, that it develops by simple chance, that the particles within it are available and that living things have been produced and developed by chance encounters of molecules, is a reduction. It is to reject, with conceptual closure, the search for the cause of the organization of the expansive Universe, as well as the search for where the available space comes from that makes it possible to generate structured elements with permanence, which will aggregate following an organizational law defined at the creation of the Universe by the generating component, and which does not cease to produce multiplications of informational components.

We have thus proposed a general model that tends to unify different scientific domains, giving each of them the same informational base that founds the material elements whose movements, positions and transformations are studied by science. In the model, we have also posited that the creation of the Universe and the sixth sense of living beings belong to the same conceptual domain, with the same concepts of organization of material aggregates and informational flows.

We have shown that the sixth sense of living beings means using the informational substrate and that it allows communications with very social consequences to organize groups. We have posited that, in all living animals, this sixth sense is the natural transposition of the communicational rule of the world, generated on the informational substrate of the Universe by communications. We have made it clear that this sixth sense is available to humans, who can use and develop it to reinforce the social nature and ethics of their societies.

We hope that applied multidisciplinary research will lead to the development and use of this sixth sense in humans, so that they can build multiple ethical associations, shared structures and organizations where humanistic communications are practiced, leading to peaceful and shared happiness. Will this just be a dream or will it be the next path in the reconstruction of our world that is currently plagued with great problems?

Appendices

Appendix 1

Binomial Distribution

In many engineering applications, events consisting of a repetition of trials can be formulated as follows: occurrence or non-occurrence, success or failure, good or bad, etc. There are only two possible solutions that represent the behavior of a random discrete variable. Moreover, if the events satisfy the additional condition of a Bernoulli sequence, in other words, if they are statistically independent and the probability of occurrence or non-occurrence remains constant, these events can be mathematically represented by a binomial distribution.

If the probability of occurrence of an event for each trial is p and the probability of its non-occurrence is (1-p), then the probability of x occurrences over a total of n trials can be determined by the mass function of a polynomial distribution:

$$P(X = x, n \mid p) = \binom{n}{x} p^x (1 - p)^{n-x} \quad x = 0, 1, 2, \ldots, n$$

where p is the distribution parameter and $\binom{n}{x} = \dfrac{n!}{x!(n-x)!}$ is the binomial coefficient, indicating the number of ways that x occurrences out of a total of n trials are possible.

Appendix 2

Geometric Distribution

The first occurrence of an event is very important in engineering. If the event occurs in a Bernoulli sequence and p is the probability of occurrence at each trial, then the probability that the event occurs at the ith trial, which implies that it has not yet occurred in previous trials, is given by the following geometric distribution:

$$P(T = t) = p(1-p)^{t-1} \quad t = 1,2,\ldots \qquad [\text{A2.1}]$$

Return period: changes in wind speed, rainfall, flooding or seismic hazards at a specific location are usually expressed in terms of return period. Let us assume an event that occurs once in a Bernoulli sequence after a period of T1 years. This event occurs again T2 years after the first occurrence, then again T3 years after the second occurrence, and so on. The recurrence time, in other words, the time between two consecutive occurrences of the same event, must follow the characteristic probability of the first occurrence, that is, a geometric distribution whose mass probability is given by equation [A2.2]. We can thus calculate the average recurrence time, which is also known as the return period:

$$T = E(T) = \sum_{t=1}^{\infty} t p_T(t) = \sum_{t=1}^{\infty} t p(1-p)^{t-1}$$

$$= p[1 + 2(1-p) + 3(1-p)^2 + 4(1-p)^3 + \cdots] \qquad [\text{A2.2}]$$

The terms in square brackets in equation [A2.3] represent an infinite series and can be described as $1/p^2$. Thus, the equation can be written in the following simplified form:

$$T = p \times \frac{1}{p^2} = \frac{1}{p} \qquad\qquad [A2.3]$$

Appendix 3

Poisson Distribution

The Poisson distribution is generally used to assess risks in industry. It takes problems in time or space into account. If we wanted to solve this kind of problem with the Bernoulli distribution, we would have to decompose the time or space interval into very small parts, given that it only takes the occurrence or absence of an event into account. Let us suppose that the average occurrence of a tornado in a certain location is v times per year, where during a period of t years, a tornado will occur on average v × t times. If the period of time is divided by n intervals, the probability of a tornado occurring will be (v × t)/n. If we take x occurrences during a period of time t in a Bernoulli sequence when n tends to infinity, then we obtain the Poisson distribution, which can be expressed as follows:

$P(x$ occurences in $t)$

$$= \lim_{n \to \infty} \binom{n}{x}\left(\frac{vt}{n}\right)^x \left(1 - \frac{vt}{n}\right)^{n-x}$$

$$= \lim_{n \to \infty} \left[\frac{n\,(n-1)}{n}\frac{}{n}\cdots\frac{(n-x+1)}{n}\cdots\frac{(vt)^x}{x!}\left(1 - \frac{vt}{n}\right)^n \left(1 - \frac{vt}{n}\right)^{-x}\right]$$

$$= \lim_{n \to \infty} \left[\frac{(vt)^x}{x!}\left(1 - \frac{vt}{n}\right)^n\right]$$

We take the limit of this equation, knowing that:

$$\lim_{n \to \infty} \left(1 - \frac{vt}{n}\right)^n = 1 - vt + \frac{(vt)^2}{2!} - \frac{(vt)^3}{3!} + \cdots = e^{-vt}$$

We can therefore show that:

$$P(x \text{ occurrences in time } t) = \frac{(v \times t)^x}{x!} e^{-v \times t}$$

Appendix 4

Exponential Distribution

If an event follows a Poisson logic, then the time T before the first occurrence of this event can be represented by the following exponential distribution:

$$P(T > t) = \frac{e^{v \times t}(v \times t)^0}{0!} = e^{-v \times t}$$

The distribution function is thus written as:

$$F_T(t) = P(T \leq t) = 1 - e^{-v \times t}$$

and the probability density function of the exponential distribution is written as:

$$f_t(t) = \frac{dF_T(t)}{dt} = v \times e^{-v \times t}, \ t \geq 0.$$

It can be shown that the average value of T is equal to $1/v$. That is, the average time of the first occurrence, or the recurrence time, or simply the return period for a Poisson model is $1/v$. It is interesting to note that when an event follows a Bernoulli model, the return period is $1/p$. It can be shown that for an event with a low occurrence v, the return periods for a Bernoulli sequence and for a Poisson model are more or less identical.

Appendix 5

Normal Distribution

A real random variable X follows a normal or Gaussian distribution of mean μ and standard deviation σ if its probability density f is given by:

$$
\begin{cases}
f(x) = \dfrac{1}{\sigma\sqrt{2\pi}}e^{-\frac{1}{2}\left(\frac{x-\mu}{\sigma}\right)^2} \\
\text{with } x \in \mathbb{R},\ \mu \in \mathbb{R},\ \sigma \in \mathbb{R}^{+*}
\end{cases}
$$

This function is determined by knowing the parameters μ and σ. We note that $X \sim \mathcal{N}(\mu,\sigma) : X$ follows the law $\mathcal{N}(\mu,\sigma)$.

The graphical representation of f(x) leads to the famous "bell curve".

Regardless of the values of the parameters, the Gaussian distribution has a maximum amplitude known as the statistical mode. The mean μ corresponds to the abscissa of this maximum amplitude. The intervals that are typically considered are [μ-σ,μ+σ], [μ-2σ,μ+2σ], [μ-3σ,μ+3σ] and [μ-4σ,μ+4σ], which respectively contain 68.3, 95.5, 99.7 and 99.99% of the considered statistical population.

The distribution function F is defined by:

$$
F(x) = \int_{-\infty}^{x} (t)dt = \int_{-\infty}^{x} \dfrac{1}{\sigma\sqrt{2\pi}}e^{-\frac{1}{2}\left(\frac{t-\mu}{\sigma}\right)^2} dt
$$

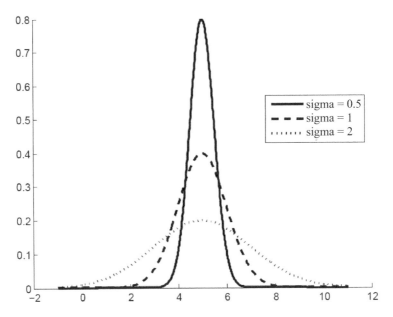

Figure A5.1. *Influence of the standard deviation at a constant mean. Probability density of the Gaussian distribution – mean of 5*

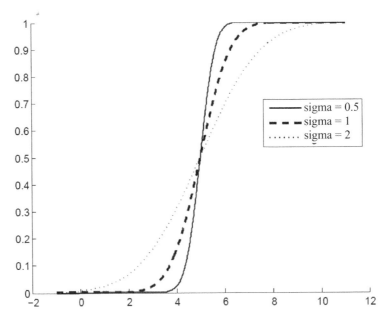

Figure A5.2. *Influence of the standard deviation at a constant mean. Distribution function of the Gaussian distribution – mean of 5*

Standard normal distribution: a Gaussian random variable X_0 of zero mean and unit standard deviation follows the standard normal distribution ($\mu=0$ and $\sigma=1$): $X_0 \sim \mathcal{N}(0,1)$. Its associated probability density is a direct result of:

$$f(x_0) = \frac{1}{\sqrt{2\pi}} e^{-\frac{x_0^2}{2}} \quad x_0 \in \mathbb{R}$$

Typically, the distribution function of the standard normal distribution is denoted as ϕ:

$$\Phi(x_0) = \int_{-\infty}^{x_0} f(t_0) dt_0 = \int_{-\infty}^{x_0} \frac{1}{\sqrt{2\pi}} e^{-\frac{1}{2}t_0^2} dt_0$$

If $X \sim \mathcal{N}(\mu,\sigma)$, then $X_0 = \dfrac{X - \mu}{\sigma} \sim \mathcal{N}(0,1)$; therefore:

$$\Phi\left(\frac{x-\mu}{\sigma}\right) = \int_{-\infty}^{\frac{x-\mu}{\sigma}} \frac{1}{\sqrt{2\pi}} e^{-\frac{1}{2}\left(\frac{t-\mu}{\sigma}\right)^2} \frac{dt}{\sigma} = \int_{-\infty}^{x_0} \frac{1}{\sigma\sqrt{2\pi}} e^{-\frac{t_0^2}{2}} dt = F(x)$$

We then obtain:

$$E(X_0) = E\left(\frac{X-\mu}{\sigma}\right) = \frac{1}{\sigma} E(X) - \frac{\mu}{\sigma} = \frac{\mu - \mu}{\sigma} = 0$$

$$V(X_0) = \frac{E(X^2) - \mu^2}{\sigma^2}$$

$$\sqrt{V(X_0)} = \frac{\sqrt{E(X^2) - \mu^2}}{|\sigma|} = \frac{\sqrt{V(X)}}{\sigma} = \frac{\sigma}{\sigma} = 1$$

with E being the mean and V being the variance of the variables considered.

The odd-order moments of the standard normal distribution are all zero. This is due to the parity of the probability density. Indeed, the moments of order n and 2k+1 are written as:

$$m_n = \int_{-\infty}^{+\infty} t^n \frac{1}{\sqrt{2\pi}} e^{-\frac{t^2}{2}} dt$$

$$m_{2k+1} = \int_{-\infty}^{+\infty} t^{2k+1} \frac{1}{\sqrt{2\pi}} e^{-\frac{t^2}{2}} dt = 0$$

and the even-order moments are written as:

$$m_{2k} = \int_{-\infty}^{+\infty} t^{2k} \frac{1}{\sqrt{2\pi}} e^{-\frac{t^2}{2}} dt$$

It is easy to show that:

$$m_{2k+2} = (2k+1) m_{2k}$$

As $m_0 = 1$, where $m_{2k} = \frac{(2k)!}{2^k k!}$.

Appendix 6

Lognormal Distribution

A random variable Y follows the lognormal distribution of parameters μ and σ if its Neperian logarithm follows the following normal distribution:

$$Y \sim \ln-\mathcal{N}(\mu,\sigma)$$
$$X = \ln(Y) \sim \mathcal{N}(\mu,\sigma)$$

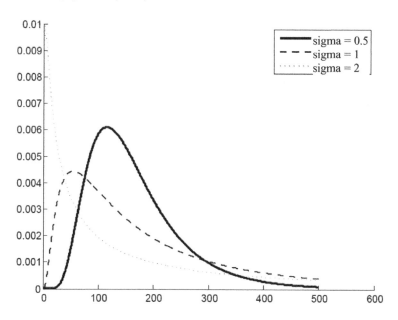

Figure A6.1. *Influence of the standard deviation at a constant mean. Probability density of the lognormal – mu = 5*

The associated probability density is:

$$f(y) = \frac{1}{\sigma y \sqrt{2\pi}} e^{-\frac{1}{2}\left(\frac{\ln(y)-\mu}{\sigma}\right)^2}$$

with $y \in \mathbb{R}^{+*}, \mu \in \mathbb{R}, \sigma \in \mathbb{R}^{+*}$.

The distribution function F is given by:

$$F(y) = \Phi\left(\frac{\ln(y) - \mu}{\sigma}\right)$$

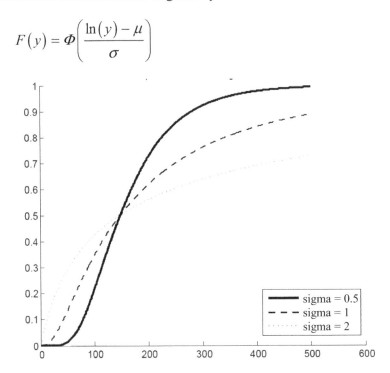

Figure A6.2. *Influence of the standard deviation at a constant mean. Distribution function of the lognormal distribution – mu = 5*

The moments of order k of the lognormal distribution are:

$$m_k = \int_0^{+\infty} y^k f(y) dy = \int_0^{+\infty} y^k \frac{1}{\sigma y \sqrt{2\pi}} e^{-\frac{1}{2}\left(\frac{\ln(y)-\mu}{\sigma}\right)^2} dy$$

After calculation, we find that $m_k = e^{k\mu + \frac{k^2 \sigma^2}{2}}$.

The mean is given by:

$$E(Y) = m_1 = e^{\mu + \frac{\sigma^2}{2}}$$

The variance is given by:

$$V(Y) = e^{2\mu + \sigma^2}\left(e^{\sigma^2} - 1\right)$$

from which the standard deviation is written as:

$$\sigma_Y = \sqrt{V(Y)} = e^{\mu + \frac{\sigma^2}{2}}\sqrt{e^{\sigma^2} - 1}$$

A property of the lognormal distribution is that, unlike the normal distribution, the mean, median and mode are not combined.

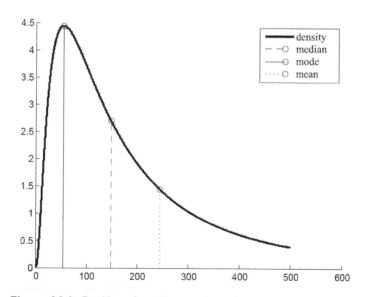

Figure A6.3. *Position of median, mode and mean, to be compared with μ. Probability density of the lognormal – mu = 5 and sigma = 1*

Appendix 7

Weibull Distribution

The Weibull distribution covers a whole family of continuous probability laws that are generically applied to the lifetimes of many types of equipment (systems, spare parts, relays, fatigue, wear and tear, etc.) (El Hami and Radi 2013a). In reliability monitoring, companies tend to use it, especially when the failure rate evolves as a power of time. When this rate is constant, the exponential distribution, which is a special form of the Weibull distribution, is used. The Weibull distribution is based on three parameters, which are:

– α: time-scale parameter;

– β: shape parameter;

– γ: location or tracking parameter.

These parameters, whose influence we will study later, allow us to modify the different periods (see Figure A7.1).

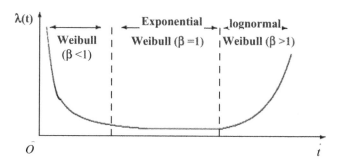

Figure A7.1. *Weibull distribution*

The different indicators are as follows:

– The damage rate. Weibull proposed the following model for the damage rate:

$$a(t) = \frac{d}{dt}\frac{(t-\gamma)^\beta}{\alpha} = \frac{\beta}{\alpha}(t-\gamma)^{\beta-1}$$

– The reliability; R(t) is the probability of good operation at time t. By definition, we obtain:

$$a(t)dt = -\frac{dR(t)}{R(t)}$$

This leads to:

$$\begin{cases} R(t) & = & \exp(-\int_0^t a(x)dx) \quad t \in [0,+\infty] \\ R(0) & = & 1 \end{cases}$$

Because:

$$\int_0^t a(x)dx = \int_0^t \frac{d}{dx}\left[\frac{(x-\gamma)^\beta}{\alpha}\right]dx = \frac{(t-\gamma)^\beta}{\alpha}$$

Ultimately, we obtain:

$$R(t) = \exp\left(-\frac{(t-\gamma)^\beta}{\alpha}\right) = \exp\left(-\left(\frac{t-\gamma}{\eta}\right)^\beta\right) \quad (\eta^\beta = \alpha)$$

We note that $\eta^\beta = \alpha$.

– The distribution function. From the reliability R(t), we directly obtain the following distribution function F(t):

$$F(t) = 1 - R(t) = 1 - \exp\left(-\left(\frac{t-\gamma}{\eta}\right)^\beta\right)$$

– The probability density. By definition, we obtain:

$$f(t) = \frac{d}{dt}F(t) = \frac{d}{dt}\left(1 - e^{\left(-\left(\frac{t-\gamma}{\eta}\right)^{\beta}\right)}\right) = \frac{\beta}{\eta}\left(\frac{t-\gamma}{\eta}\right)^{\beta-1}.e^{\left(-\left(\frac{t-\gamma}{\eta}\right)^{\beta}\right)}$$

– The instant failure rate. By definition, we obtain:

$$\lambda(t) = \frac{f(t)}{R(t)} = \frac{\dfrac{\beta}{\eta}.\left(\dfrac{t-\gamma}{\eta}\right)^{\beta-1}.e^{\left(-\left(\frac{t-\gamma}{\eta}\right)^{\beta}\right)}}{e^{-\left(\frac{t-\gamma}{\eta}\right)^{\beta}}}$$

$\lambda(t)$ is a reliability estimator;

– The lifetime; the lifetime is determined from R(t). We obtain:

$$R(t) = e^{-\left(\frac{t-\gamma}{\eta}\right)^{\beta}}$$

where:

$$\ln(R(t)) = -\left(\frac{t-\gamma}{\eta}\right)^{\beta} \Rightarrow \ln\left(\frac{1}{R(t)}\right) = \left(\frac{t-\gamma}{\eta}\right)^{\beta} \Rightarrow \ln\left(\frac{1}{R(t)}\right)^{\frac{1}{\beta}} = \frac{t-\gamma}{\eta}$$

Ultimately, we obtain:

$$t = \eta.\ln\left(\frac{1}{R(t)}\right)^{\frac{1}{\beta}} + \gamma$$

– The mean. The expression of the mean is:

$$\bar{x} = \alpha^{\frac{1}{\beta}} . \frac{1}{\beta} . \Gamma\left(\frac{1}{\beta}\right) = \eta.\frac{1}{\beta}\Gamma\left(\frac{1}{\beta}\right)$$

We recall that the function is defined by $\Gamma(x) = \int_0^{+\infty} t^{x-1} e^{-t} dt$. The function Γ is described in the statistical tables.

– The variance. The expression of the variance is given from the standard deviation σ by:

$$\sigma^2 = \eta^2 \left(\Gamma\left(\frac{2}{\beta} + 1\right) - \Gamma\left(\frac{1}{\beta} - 1\right)^2 \right)$$

Appendix 8

Pareto Distribution

The Pareto variable, of parameters α and t_0 ($t_0 \geq 0$), is the random variable t whose distribution function is given by:

$$
\begin{cases}
F(t) & = & 0 & \text{si } t < t_0 \\
F(t) & = & 1 - \left(\dfrac{t_0}{t}\right)^{\alpha} & \text{si } t \geq t_0
\end{cases}
$$

with:

$P\{t \text{ lifetime} < t\} = F(t)$

$P\{t \text{ lifetime} \geq t\} = R(t) = 1 - F(t)$

The density of this variable is given by:

$$
f(t) = \frac{dF(t)}{dt} = \frac{d(-t_0^{\alpha} \times t^{-\alpha})}{dt}
$$

So:

$$
f(t) = \frac{\alpha}{t_0} \cdot \left(\frac{t_0}{t}\right)^{\alpha+1}
$$

There are two characteristics of the Pareto distribution: the quantiles and the moments. We thus have:

1) Quantiles:

$$F(t_p) = 1 - \left(\frac{t_0}{t_p}\right)^\alpha = p$$

Therefore:

$$t_p = \frac{t_0}{(1-p)^{1/\alpha}}$$

The median is given by $M_e = t_{0,5} = t_0 . 2^{1/\alpha}$.

2) Moments: the uncentered moment of order r is m_r, which is calculated as follows:

$$m_r = \int_{t_0}^{+\infty} f(t)t^r \, dt = \frac{\alpha t_0^r}{\alpha - r} \quad \text{si } \alpha < r$$

The mean $\bar{t}$ exists only if $\alpha > 1$.

$$\bar{t} = m_1 = \frac{\alpha}{\alpha - 1}.t_0$$

The variance V exists only if $\alpha > 2$.

$$V = m_2 - m_1^2 = \frac{\alpha . t_0^2}{\alpha - 2} - (\frac{\alpha}{\alpha - 1} t_0)^2$$

So:

$$V = \frac{\alpha . t_0^2}{(\alpha - 2)(\alpha - 1)^2}$$

The standard deviation σ (t) is:

$$\sigma(t) = \frac{t_0}{\alpha - 1} \cdot \sqrt{\frac{\alpha}{\alpha - 2}}$$

Appendix 9

Distribution of Extreme Values

In many engineering applications, the extreme values of random variables have an important meaning. For example, the highest or lowest value may indicate a particular pattern. Take the example of wind speed, which is continuously recorded at airports or weather stations. Evidently, the very large amount of data cannot be used as it is in the mathematical models. However, the maximum wind speed per hour, day, month, year or any other period can be used for this purpose. Usually, the information about the maximum wind speed over a year is considered in the applications. Thus, for each batch of data for a year, the maximum wind speed is recorded. If the data is recorded for many years, the wind speed model can be established statistically to ensure that it will not exceed the lifetime for which the structure was designed.

If the wind speed model has a return period of 50 years, then the probability that the wind speed will exceed the design speed by one day in one year is 1/50=0.02. Earthquakes, floods and other climatic events can thus be determined. In all these cases, the maximum value of a random variable during a certain period is important to consider. In some cases, the minimum value is also important for the analysis. For example, when a large number of devices are manufactured, such as computers or cars, their minimum service life is an important point to consider for consumers. Thus, accelerated life tests can be used to build a probability of life distribution. In this way, the minimum service life can be established so that it does not fall below a value that is deemed acceptable. In constructing an extreme value distribution, it is necessary to establish an underlying random variable with a particular distribution. If different sets of samples are obtained (through physical or numerical experiments), we can select the extreme values for each set of samples, and thus construct a particular distribution for the extreme values.

Concept of extreme value distribution: let X be a random variable with an unknown distribution function. If there are n samples for the random variable X, the limit values for the samples, such as the minimum value Y_1 or the maximum value Y_n, could be of interest. Y_1 and Y_n can be defined as follows:

$$Y_n = \max(X_1, X_2, \ldots, X_n) \quad et \quad Y_1 = \min(X_1, X_2, \ldots, X_n) \qquad [A9.1]$$

If different sets of samples of the same size n are obtained for X, each set will have different maximum and minimum values. Using all these sets, distribution functions for the minimum and maximum values can be constructed. The cumulative distribution function (CDF) of the largest Y_n value is:

$$F_{Y_n}(y) = P(Y_n \leq y) = P(X_1 \leq y, X_2 \leq y, \ldots, X_n \leq y) \qquad [A9.2]$$

For uniformly distributed and statistically equivalent variables X_i, the previous equation becomes:

$$F_{Y_n}(y) = [F_X(y)]^n \qquad [A9.3]$$

Similarly, the CDF of the smallest Y_1 value is:

$$P(Y_1 > y) = P(X_1 > y, X_2 > y, \ldots, X_n > y) = 1 - F_{Y_1}(y) \qquad [A9.4]$$

For uniformly distributed and statistically independent X_i, equation [A9.4] becomes:

$$F_{Y_1}(y) = 1 - [1 - F_X(y)]^n \qquad [A9.5]$$

The main idea of this formula is that if the largest value Y_n is less than y, then all the values in the sample (X_1, X_2, etc.) must also be less than y, as must the smallest value Y_1.

Appendix 10

Asymptotic Distributions

The sample size grows very large and tends to infinity, and the distribution of the largest or smallest values may tend asymptotically to a mathematical distribution function in some cases when the samples are identically distributed and statistically independent. Some of these asymptotic distributions have a wide range of applications.

El Hami and Radi (2013b) listed three types of so-called asymptotic extreme value distributions as type I, type II and type III. Type I extreme value distributions are used extensively in mechanical reliability applications.

The distribution of maxima in sets of samples drawn from a population, following a normal distribution, converges asymptotically to this distribution. This type of distribution is used to model climatic phenomena such as wind or water level variations.

Type II distributions are also used to create models of environmental phenomena, such as earthquakes, and can be derived from the sample sets of a lognormal distribution.

Type III distributions, also known as Weibull distributions in the case of smaller values, can be obtained by the convergence of most of the distributions that have a lower bound. It is often used to describe the mechanical strength of a material or the good operating time of an electronic or mechanical device.

Extreme value distributions can be treated in the same way as other distributions, that is, defined in terms of their probability density function or distribution function, as well as the associated parameters. In most cases, the parameters can be estimated from information about the mean, variance or coefficient of variation of the random variable. Once an extreme value distribution is exclusively described, the probabilistic information can be extracted using the methods presented previously.

– Type I distribution: the cumulative distribution function of the asymmetric type I form, the largest value distribution, also known as the Gumbel distribution, can be expressed as follows:

$$F_{Y_n}(y_n) = \exp\left[-e^{\alpha_n(y_n - u_n)}\right]$$

where u_n is the largest characteristic value of the initial variable X and α_n is the inverse measure of the dispersion of the largest value of X. It can be shown that the probability distribution function is:

$$f_Y(y) = \alpha e^{\alpha_n(y_n - u_n)} \exp\left[-e^{\alpha_n(y_n - u_n)}\right], \quad -\infty < y < +\infty$$

U_n and α_n are related to the mean and standard deviation of the extreme variable Y_n:

$$\alpha_n = \frac{1}{\sqrt{6}}\left(\frac{\pi}{\sigma_{Y_n}}\right) \quad \text{and} \quad u_n = \mu_{Y_n} - \frac{0.5772}{\alpha_n}$$

Type I distribution for the extreme value of the largest value is commonly used for modeling natural phenomena. It is a very important variable in the design of flood controls, drinking water consumption and irrigation systems.

For the smallest value of a variable X, the corresponding asymptotic type I form for the cumulative distribution function is:

$$F_{Y_1}(y_1) = 1 - \exp\left[-e^{\alpha_1(y_1 - u_1)}\right]$$

and the probability density function is:

$$f_{Y_1} = \alpha_1 e^{\alpha_1(y_1-u_1)} \exp\left[-e^{\alpha_1(y_1-u_1)}\right], \quad -\infty < y_1 < +\infty$$

– Type II distribution: the distribution function for a type II distribution of a maximum value, also known as the Fréchet distribution, can be expressed as follows:

$$F_{Y_n}(Y_n) = \exp\left[-\left(\frac{v_n}{y_n}\right)^k\right]$$

The corresponding probability density function is:

$$f_{Y_n}(y_n) = \frac{k}{v_n}\left(\frac{v_n}{y_n}\right)^{k+1} \exp\left[-\left(\frac{v_n}{y_n}\right)^k\right], \quad y_n \leq 0, k > 2$$

where v_n and k are the parameters of the distribution, v_n is the maximum value of the underlying variable X and k, the shape parameter, is a measure of the dispersion.

The asymptotic form of a type II distribution is obtained when n tends to infinity from an initial distribution that has a polynomial tangent towards the extreme value. We can note the difference between this feature and type I, which converges from an exponential tangent. The type II distribution requires a polynomial direction, and therefore a lognormal distribution converges to an asymptotic type II form for maximum values.

– Type III distribution: type III asymptotic distributions represent a distribution for knowing the upper or lower bounds of distributions that have finite limits. The cumulative distribution function for the asymptotic type III can be described as follows:

$$F_{Y_n}(y_n) = \exp\left[-\left(\frac{\omega - y_n}{\omega - w_n}\right)^k\right]$$

and the probability density function:

$$f_{Y_n}(y_n)\frac{k}{\omega-w_n}\left(\frac{\omega-y_n}{\omega-w_n}\right)^{k-1}\exp\left[-\left(\frac{\omega-y_n}{\omega-w_n}\right)^k\right], y_n \leq \omega$$

where ω is the upper bound of the initial distribution; in other words, $F_X(w_n)=1-1/n$. w_n and k are the parameters of the distribution. w_n is the largest characteristic value of X and is defined by $F_X(w_n)=1-1/n$, and k is the shape parameter. The mean and variance of Y_n are related to parameters w_n and k in the following way:

$$\mu_{Y_n} = \omega - (\omega - w_n)\Gamma\left(1+\frac{1}{k}\right)$$

and:

$$\sigma_{Y_n}^2 = Var(\omega - Y_n) = (\omega - w_n)^2\left[\Gamma\left(1+\frac{2}{k}\right)-\Gamma^2\left(1+\frac{1}{k}\right)\right]$$

Using the two previous equations, we can show that:

$$1+\left(\frac{\sigma_{Y_n}}{\omega-\mu_{Y_n}}\right)^2 = \frac{\Gamma\left(1+\frac{2}{k}\right)}{\Gamma^2\left(1+\frac{1}{k}\right)}$$

The type III cumulative distribution function for the smallest value is defined as follows:

$$F_{Y_1}(y_1) = \exp\left[-\left(\frac{y_1-\epsilon}{w_1-\epsilon}\right)^k\right]$$

and the probability density function is written as:

$$f_{Y_1}(y_1) \frac{k}{w_1 - \epsilon}\left(\frac{y_1 - \epsilon}{w_1 - \epsilon}\right)^{k-1} \exp\left[-\left(\frac{y_1 - \epsilon}{w_1 - \epsilon}\right)^k\right], y_1 \geq \epsilon$$

The mean and variance of Y_1 are related to parameters w_1 and k in the following way:

$$\mu_{Y_1} = \epsilon + (w_n - \epsilon)\Gamma\left(1 + \frac{1}{k}\right)$$

and:

$$\sigma_{Y_1}^2 = Var(Y_1 - \epsilon) = (w_1 - \epsilon)^2\left[\Gamma\left(1 + \frac{2}{k}\right) - \Gamma^2\left(1 + \frac{1}{k}\right)\right]$$

Using the two previous equations, we can show that:

$$1 + \left(\frac{\sigma_{Y_1}}{\mu_{Y_1} - \epsilon}\right)^2 = \frac{\Gamma\left(1 + \frac{2}{k}\right)}{\Gamma^2\left(1 + \frac{1}{k}\right)}$$

References

Bomsel, M.-C. (2006). *Leur sixième sens : les animaux sont-ils plus "sensés" que nous* ? Michel Lafon.

Bouyekhf, R., El Moudni, A., El Hami, A., Zerhouni, N., Ferney, M. (1996). Reduced order modelling of two-time-scale discrete non-linear systems. *Journal of the Franklin Institute*, 333(4), 499–512.

Bouyekhf, R., El Hami, A., El Moudni, A. (2001). Optimal control of a particular class of singularly perturbed nonlinear discrete-time systems. *IEEE Transactions on Automatic Control*, 46(7), 1097–1101.

Brohm, J.-M. (2017). *Ontologies du corps*. Presses universitaires de Paris Nanterre.

Campagne, J.C. (2005). Systèmes multi-agents et morphologie. PhD Thesis, Université Pierre et Marie Curie.

Cardon, A. (2018). *Beyond Artificial Intelligence: From Human Consciousness to Artificial Consciousness*. ISTE Ltd, London, and John Wiley & Sons, New York.

Cardon, A. (2022). *Information Organization of The Universe and Living Things: Generation of Space, Quantum and Molecular Elements, Coactive Generation of Living Organisms and Multiagent Model*. ISTE Ltd, London, and John Wiley & Sons, New York.

Cardon, A. and Itmi, M. (2016). *New Autonomous Systems*. ISTE Ltd, London, and John Wiley & Sons, New York.

Damasio, A. and Bechara, A. (2005). The somatic maker hypothesis: A neural theory of economic decision, in games and economic. *Behavior*, 52(2), 336–372.

El Hami, A. and Radi, B. (2011). Comparison study of different reliability-based design optimization approaches. *Advanced Materials Research*, 274, 113–121.

El Hami, A. and Radi, B. (2013a). *Incertitude et optimisation et fiabilité des structures*. Hermès-Lavoisier.

El Hami, A. and Radi, B. (2013b). *Uncertainty and Optimization in Structural Mechanics*. ISTE Ltd, London, and John Wiley & Sons, New York.

El Hami, A., Lallement, G., Minotti, P., Cogan, S. (1993). Methods that combine finite group theory with component mode synthesis in the analysis of repetitive structures. *Computers and Structures*, 48(6), 975–982.

Feldmeyer, J.-J. (2002). *Cerveau et pensée, la conquête des neurosciences*. Georg.

Ferber, J. (1999). *Multi-Agent System: An Introduction to Distributed Artificial Intelligence Harlow*. Addison Wesley Longman.

Freud, S. (1966). *The Complete Psychological Works of S. Freud*, translated by J. Strachey. The Hogarth Press.

Gallese, V. and Massimo, A. (2015). *Psychothérapie et neuroscience : une nouvelle alliance*. Faber.

Hardy, C. (2012). *La prédiction de Jung : la métamorphose de la Terre*. Dervy-Livres.

Haroche, C. and Aubert, N. (2011). *Les tyrannies de la visibilité : être visible pour exister*. Érès.

Hiernaux, Q. (2020). *Du comportement végétal à l'intelligence des plantes*. Quae.

Jamshidi, M. (2008). *Systems of Systems Engineering: Principles and Applications*. CRC Press.

Jannoun, M., Aoues, Y., Pagnacco, E., Pougnet, P., El-Hami, A. (2017). Probabilistic fatigue damage estimation of embedded electronic solder joints under random vibration. *Microelectronics Reliability*, 78, 249–257.

Kharmanda, G., Ibrahim, M.-H., Abo Al-kheer, A., Guerin, F., El-Hami, A. (2014). Reliability-based design optimization of shank chisel plough using optimum safety factor strategy. *Computers and Electronics in Agriculture*, 109, 162–171.

Maharishi, M.Y. (1976). *La science de l'être et l'art de vivre*. Robert Laffont.

Maier, M. (1999). Architecting principles for systems-of-systems. *Systems Engineering*, 2(1), 1.

Marchais, P. and Cardon, A. (2010). *Troubles mentaux et interprétations informatiques*. L'Harmattan.

Marchais, P. and Cardon, A. (2015). Nouvelle perspective en psychiatrie. De la globalité psychique à la multiplicité des troubles mentaux. *Annales Médico-Psychologiques*, 174(2), 85–92.

Mataric, M. (1997). Behavior based control: Examples from navigation, learning and group behavior. *Journal of Experimental and Theorical Artificial Intelligence*, 9, 323–336.

Moore, G.E. (1965). Cramming more components onto integrated circuits. *Electronics*, 38(8) [Online]. Available at: https://web.archive.org/web/20190327213847/ https://newsroom.intel.com/wp-content/uploads/sites/11/2018/05/moores-law-electronics.pdf.

Newell, A. (1982). The knowledge level. *Artificiel Intelligence*, 18, 81–127.

Newman, M.H. and Geoffrey Jefferson, R.B. (2004). Braithwaite. *Can Automatic Calculating Machines be Said to Think?* [Online]. Available at: http://www.turingarchive.org/brows.

Ören, T., Kazemifard, M., Noori, F. (2015). Agents with four categories of understanding abilities and their role in two-stage (deep) emotional intelligence simulation. *International Journal of Modeling, Simulation and Scientific Computing*, 6(3) [Online]. Available at: https://doi.org/10.1142/ S1793962315400036.

Sarano, F. (2017). *Le retour de Moby Dick, ou ce que les cachalots nous enseignent sur les océans et les hommes*. Actes Sud.

Serra, J. (1982). *Image Analysis and Mathematical Morphology*. Academic Press.

Shannon, C.E. (1948). A mathematical theory of communication. *Bell System Technical Journal*. 27(3), 379–423 [Online]. Available at: https://ieeexplore.ieee.org/ document/6773024.

Thom, R. (1972). *Stabilité structurelle et morphogenèse*. W.A. Benjamin Inc.

Varela, F. (1989). *Autonomie et connaissance. Essai sur le vivant*. Le Seuil.

Verstrynge, J. (2010). *Practical JXTA II. Craking the P2P puzzle*. Lulu Enterprises.

de Wailly, P. (2009). *Le sixième sens des animaux*. J'ai Lu.

Wooldridge, M. and Jennings, N.R. (1994). *Agent Theories, Architectures and Languages: A Survey*. Springer.

Zeigler, B.P. and Sarjoughian, H.S. (2013). *Guide to Modeling and Simulation of Systems of Systems*. Springer.

Index

A, B, C

activity component, 64, 65, 68, 69, 71, 72, 74–77
agent
    design, 57
    notion
        strong, 11
        weak, 10
    regulation, 57
apprehension of the near future, 152
artificial
    corporeality, 8
    organ, 8
brains
    formation of, 97
central rule of organizational law in living things, 84
creation
    of organisms, 84
    of sexual reproduction, 89

E, F, G, H

emotional processing center, 58, 110, 117, 121, 123
five usual senses, 110, 125, 128, 137, 141, 142

generating element of the Universe, 64
Homo sapiens, 100, 101
human beings closed in on themselves, 146
hypnosis, 147

I, L

information
    comprehension of the transmitted information, 135
    field of a component, 72
    generative, 71, 72, 75
    transmitted by the informational link, 134
informational
    bifurcation, 82
    component, 65, 68, 70, 71, 73–75, 77
    comprehension (sense of), 125, 129
    energy, 64–66, 68–71, 73, 74, 77, 94, 104
    geographic map, 138
    substrate of the Universe, 39, 156

informational envelope(s), 71–74,
    76–79, 81, 85, 87, 88, 90–95, 99,
    101, 104, 105, 123–125, 128–139,
    142–145, 147, 148, 152
  of the brain, 135, 137
  unification between the
      informational envelope and
      external membrane, 88
informational link(s)
  activation, 132, 133, 137
  information transmitted by the, 134
  permanent, 143
layer
  organizational, 59, 60, 121, 122
  systemic, 112, 114, 116, 117
life span of a living organism, 87

M

magnetism, 125, 148
  power of, 147, 148
meditation, 119, 147, 149, 150, 153
morphological pattern(s), 93, 96, 103,
    109, 130, 132, 144, 149
  dominant, 96, 130, 132, 144
  influence of, 95

O, P

organizational
  attractors, 99
  law, 63, 64, 66–70, 74, 77–79,
      81–90, 92–94, 96, 97, 99,
      101–103, 129, 134, 136, 150,
      154, 155
  memory, 59, 60, 110, 111, 121–123
preconscious, 58–60, 110–112, 116,
    117, 119, 121–123
premonition, 151

R, S, T

regulator(s), 59, 93, 112–118,
    121–124, 127, 133, 146, 150
  organizational immersion, 123,
      124, 131
representation
  felt mental, 108
  ideal, 123
  informational mental, 141
requestor and receiver, 134, 138
significant neuronal aggregates, 107
sixth sense
  cerebral domain, 135–137,
      139–141, 144–146
  common use, 142
  two cases of use of the, 130
system
  complex, 3
  generating, 64, 65
  interpretation, 9
  multi-agent, 12
  of systems, 3
thought system, 110

Other titles from

in

Science, Society and New Technologies

2023

ELAMÉ Esoh
The Sustainable City in Africa Facing the Challenge of Liquid Sanitation
(Territory Development Set – Volume 2)

JURCZENKO Emmanuel
Climate Investing: New Strategies and Implementation Challenges

2022

AIT HADDOU Hassan, TOUBANOS Dimitri, VILLIEN Philippe
Ecological Transition in Education and Research

CARDON Alain
Information Organization of The Universe and Living Things: Generation of
Space, Quantum and Molecular Elements, Coactive Generation of Living
Organisms and Multiagent Model
(Digital Science Set – Volume 3)

CAULI Marie, FAVIER Laurence, JEANNAS Jean-Yves
Digital Dictionary

DAVERNE-BAILLY Carole, WITTORSKI Richard
Research Methodology in Education and Training: Postures, Practices and Forms
(Education Set – Volume 12)

ELAMÉ Esoh
Sustainable Intercultural Urbanism at the Service of the African City of Tomorrow
(Territory Development Set – Volume 1)

FLEURET Sébastien
A Back and Forth Between Tourism and Health: From Medical Tourism to Global Health
(Tourism and Mobility Systems Set – Volume 5)

KAMPELIS Nikos, KOLOKOTSA Denia
Smart Zero-energy Buildings and Communities for Smart Grids
(Engineering, Energy and Architecture Set – Volume 9)

2021

BARDIOT Clarisse
Performing Arts and Digital Humanities: From Traces to Data
(Traces Set – Volume 5)

BENSRHAIR Abdelaziz, BAPIN Thierry
From AI to Autonomous and Connected Vehicles: Advanced Driver-Assistance Systems (ADAS)
(Digital Science Set – Volume 2)

DOUAY Nicolas, MINJA Michael
Urban Planning for Transitions

GALINON-MÉLÉNEC Béatrice
The Trace Odyssey 1: A Journey Beyond Appearances
(Traces Set – Volume 4)

HENRY Antoine
Platform and Collective Intelligence: Digital Ecosystem of Organizations

LE LAY Stéphane, SAVIGNAC Emmanuelle, LÉNEL Pierre, FRANCES Jean
The Gamification of Society
(Research, Innovative Theories and Methods in SSH Set – Volume 2)

RADI Bouchaïb, EL HAMI Abdelkhalak
Optimizations and Programming: Linear, Non-linear, Dynamic, Stochastic and Applications with Matlab
(Digital Science Set – Volume 1)

2020

BARNOUIN Jacques
The World's Construction Mechanism: Trajectories, Imbalances and the Future of Societies
(Interdisciplinarity between Biological Sciences and Social Sciences Set – Volume 4)

ÇAĞLAR Nur, CURULLI Irene G., SIPAHIOĞLU Işıl Ruhi, MAVROMATIDIS Lazaros
Thresholds in Architectural Education (Engineering, Energy and Architecture Set – Volume 7)

DUBOIS Michel J.F.
Humans in the Making: In the Beginning was Technique
(Social Interdisciplinarity Set – Volume 4)

ETCHEVERRIA Olivier
The Restaurant, A Geographical Approach: From Invention to Gourmet Tourist Destinations
(Tourism and Mobility Systems Set – Volume 3)

GREFE GWENAËLLE, PEYRAT-GUILLARD DOMINIQUE
Shapes of Tourism Employment: HRM in the Worlds of Hotels and Air Transport (Tourism and Mobility Systems Set – Volume 4)

JEANNERET Yves
The Trace Factory
(Traces Set – Volume 3)

KATSAFADOS Petros, MAVROMATIDIS Elias, SPYROU Christos
Numerical Weather Prediction and Data Assimilation (Engineering, Energy and Architecture Set – Volume 6)

KOLOKOTSA Denia, KAMPELIS Nikos
Smart Buildings, Smart Communities and Demand Response (Engineering, Energy and Architecture Set – Volume 8)

MARTI Caroline
Cultural Mediations of Brands: Unadvertization and Quest for Authority (Communication Approaches to Commercial Mediation Set – Volume 1)

MAVROMATIDIS Lazaros E.
Climatic Heterotopias as Spaces of Inclusion: Sew Up the Urban Fabric (Research in Architectural Education Set – Volume 1)

MOURATIDOU Eleni
Re-presentation Policies of the Fashion Industry: Discourse, Apparatus and Power (Communication Approaches to Commercial Mediation Set – Volume 2)

SCHMITT Daniel, THÉBAULT Marine, BURCZYKOWSKI Ludovic
Image Beyond the Screen: Projection Mapping

VIOLIER Philippe, with the collaboration of TAUNAY Benjamin
The Tourist Places of the World
(Tourism and Mobility Systems Set – Volume 2)

2019

BRIANÇON Muriel
The Meaning of Otherness in Education: Stakes, Forms, Process, Thoughts and Transfers
(Education Set – Volume 3)

DESCHAMPS Jacqueline
Mediation: A Concept for Information and Communication Sciences
(Concepts to Conceive 21st Century Society Set – Volume 1)

DOUSSET Laurent, PARK Sejin, GUILLE-ESCURET Georges
Kinship, Ecology and History: Renewal of Conjunctures
(Interdisciplinarity between Biological Sciences and Social Sciences Set –
Volume 3)

DUPONT Olivier
Power
(Concepts to Conceive 21ˢᵗ Century Society Set – Volume 2)

FERRARATO Coline
Prospective Philosophy of Software: A Simondonian Study

GUAAYBESS Tourya
The Media in Arab Countries: From Development Theories to Cooperation
Policies

HAGÈGE Hélène
Education for Responsibility
(Education Set – Volume 4)

LARDELLIER Pascal
The Ritual Institution of Society
(Traces Set – Volume 2)

LARROCHE Valérie
The Dispositif
(Concepts to Conceive 21ˢᵗ Century Society Set – Volume 3)

LATERRASSE Jean
Transport and Town Planning: The City in Search of Sustainable
Development

LENOIR Virgil Cristian
Ethically Structured Processes
(Innovation and Responsibility Set – Volume 4)

LOPEZ Fanny, PELLEGRINO Margot, COUTARD Olivier
Local Energy Autonomy: Spaces, Scales, Politics
(Urban Engineering Set – Volume 1)

METZGER Jean-Paul
Discourse: A Concept for Information and Communication Sciences
(Concepts to Conceive 21st Century Society Set – Volume 4)

MICHA Irini, VAIOU Dina
Alternative Takes to the City
(Engineering, Energy and Architecture Set – Volume 5)

PÉLISSIER Chrysta
Learner Support in Online Learning Environments

PIETTE Albert
Theoretical Anthropology or How to Observe a Human Being
(Research, Innovative Theories and Methods in SSH Set – Volume 1)

PIRIOU Jérôme
The Tourist Region: A Co-Construction of Tourism Stakeholders
(Tourism and Mobility Systems Set – Volume 1)

PUMAIN Denise
Geographical Modeling: Cities and Territories
(Modeling Methodologies in Social Sciences Set – Volume 2)

WALDECK Roger
Methods and Interdisciplinarity
(Modeling Methodologies in Social Sciences Set – Volume 1)

2018

BARTHES Angela, CHAMPOLLION Pierre, ALPE Yves
Evolutions of the Complex Relationship Between Education and Territories
(Education Set – Volume 1)

BÉRANGER Jérôme
The Algorithmic Code of Ethics: Ethics at the Bedside of the Digital Revolution
(Technological Prospects and Social Applications Set – Volume 2)

DUGUÉ Bernard
Time, Emergences and Communications
(Engineering, Energy and Architecture Set – Volume 4)

GEORGANTOPOULOU Christina G., GEORGANTOPOULOS George A.
Fluid Mechanics in Channel, Pipe and Aerodynamic Design Geometries 1
(Engineering, Energy and Architecture Set – Volume 2)

GEORGANTOPOULOU Christina G., GEORGANTOPOULOS George A.
Fluid Mechanics in Channel, Pipe and Aerodynamic Design Geometries 2
(Engineering, Energy and Architecture Set – Volume 3)

GUILLE-ESCURET Georges
Social Structures and Natural Systems: Is a Scientific Assemblage
Workable?
(Social Interdisciplinarity Set – Volume 2)

LARINI Michel, BARTHES Angela
Quantitative and Statistical Data in Education: From Data Collection to
Data Processing
(Education Set – Volume 2)

LELEU-MERVIEL Sylvie
Informational Tracking
(Traces Set – Volume 1)

SALGUES Bruno
Society 5.0: Industry of the Future, Technologies, Methods and Tools
(Technological Prospects and Social Applications Set – Volume 1)

TRESTINI Marc
Modeling of Next Generation Digital Learning Environments: Complex
Systems Theory

2017

ANICHINI Giulia, CARRARO Flavia, GESLIN Philippe,
GUILLE-ESCURET Georges
Technicity vs Scientificity – Complementarities and Rivalries
(Interdisciplinarity between Biological Sciences and Social Sciences Set –
Volume 2)

DUGUÉ Bernard
Information and the World Stage – From Philosophy to Science,
the World of Forms and Communications
(Engineering, Energy and Architecture Set – Volume 1)

GESLIN Philippe
Inside Anthropotechnology – User and Culture Centered Experience
(Social Interdisciplinarity Set – Volume 1)

GORIA Stéphane
Methods and Tools for Creative Competitive Intelligence

KEMBELLEC Gérald, BROUDOUS EVELYNE
Reading and Writing Knowledge in Scientific Communities: Digital
Humanities and Knowledge Construction

MAESSCHALCK Marc
Reflexive Governance for Research and Innovative Knowledge
(Responsible Research and Innovation Set - Volume 6)

PARK Sejin, GUILLE-ESCURET Georges
Sociobiology vs Socioecology: Consequences of an Unraveling Debate
(Interdisciplinarity between Biological Sciences and Social Sciences Set –
Volume 1)

PELLÉ Sophie
Business, Innovation and Responsibility
(Responsible Research and Innovation Set – Volume 7)

2016

BRONNER Gérald
Belief and Misbelief Asymmetry on the Internet

EL FALLAH SEGHROUCHNI Amal, ISHIKAWA Fuyuki, HÉRAULT Laurent,
TOKUDA Hideyuki
Enablers for Smart Cities

GIANNI Robert
Responsibility and Freedom
(Responsible Research and Innovation Set – Volume 2)

GRUNWALD Armin
The Hermeneutic Side of Responsible Research and Innovation
(Responsible Research and Innovation Set – Volume 5)

LAGRAÑA Fernando
E-mail and Behavioral Changes: Uses and Misuses of Electronic
Communications

LENOIR Virgil Cristian
Ethical Efficiency: Responsibility and Contingency
(Responsible Research and Innovation Set – Volume 1)

MAESSCHALCK Marc
Reflexive Governance for Research and Innovative Knowledge
(Responsible Research and Innovation Set – Volume 6)

PELLÉ Sophie, REBER Bernard
From Ethical Review to Responsible Research and Innovation
(Responsible Research and Innovation Set – Volume 3)

REBER Bernard
Precautionary Principle, Pluralism and Deliberation: Sciences and Ethics
(Responsible Research and Innovation Set – Volume 4)

VENTRE Daniel
Information Warfare – 2nd edition

Printed and bound by CPI Group (UK) Ltd, Croydon, CR0 4YY

27/10/2024